TOXIN NATION

The Poisoning of Our Air, Water, Food, and Bodies

Marie D. Jones

ABOUT THE AUTHOR

Marie D. Jones is a fully trained disaster response/preparedness member of Community Emergency Response Teams (CERT) through FEMA and the Department of Homeland Security, and she is a licensed ham radio operator (KI6YES). She is the author of over 15 nonfiction books on cutting-edge science, the paranormal, conspiracies, ancient knowledge, and unknown mysteries, including Visible Ink Press' *Celebrity Ghosts and Notorious Hauntings; Demons, the Devil, and Fallen Angels; The Disaster Survival Guide: How to Prepare For and Survive Floods, Fires, Earthquakes and More; Earth Magic: Your Complete Guide to Natural Spells, Potions, Plants, Herbs, Witchcraft, and More;* and *The New Witch: Your Guide to Modern Witchcraft, Wicca, Spells, Potions, Magic, and More.* She is also the author of *PSIence: How New Discoveries in Quantum Physics and New Science May Explain the Existence of Paranormal Phenomena; 2013: End of Days or a New Beginning; Supervolcano: The Catastrophic Event That Changed the Course of Human History;* and *The Grid: Exploring the Hidden Infrastructure of Reality.* She is a regular contributor to *New Dawn Magazine, FATE, Paranoia Magazine,* and other periodicals. Jones has been interviewed on over a thousand radio shows worldwide, including *Coast-to-Coast AM.* She makes her home in San Marcos, California, and is the mom to one very brilliant son, Max.

TOXIN NATION

The Poisoning of Our Air, Water, Food, and Bodies

Marie D. Jones

MORE VISIBLE INK PRESS BOOKS BY MARIE D. JONES

Celebrity Ghosts and Notorious Hauntings; Demons, the Devil, and Fallen Angels
ISBN: 978-1-57859-689-8

Demons, the Devil, and Fallen Angels
ISBN: 978-1-57859-613-3

Disinformation and You: Identify Propaganda and Manipulation
ISBN: 978-1-57859-740-6

The Disaster Survival Guide: How to Prepare For and Survive Floods, Fires, Earthquakes and More
ISBN: 978-1-57859-673-7

Earth Magic: Your Complete Guide to Natural Spells, Potions, Plants, Herbs, Witchcraft, and More
ISBN: 978-1-57859-697-3

The New Witch: Your Guide to Modern Witchcraft, Wicca, Spells, Potions, Magic, and More
ISBN: 978-1-57859-716-1

ALSO FROM VISIBLE INK PRESS

Ancient Gods: Lost Histories, Hidden Truths, and the Conspiracy of Silence
by Jim Willis
ISBN: 978-1-57859-614-0

Area 51: The Revealing Truth of UFOs, Secret Aircraft, Cover-Ups, & Conspiracies
by Nick Redfern
ISBN: 978-1-57859-672-0

Assassinations: The Plots, Politics, and Powers behind History-changing Murders
by Nick Redfern
ISBN: 978-1-57859-690-4

Conspiracies and Secret Societies: The Complete Dossier, 2nd edition
by Brad Steiger and Sherry Hansen Steiger
ISBN: 978-1-57859-368-2

Control: MKUltra, Chemtrails, and the Conspiracy to Suppress the Masses
by Nick Redfern
ISBN: 978-1-57859-638-6

Cover-Ups & Secrets: The Complete Guide to Government Conspiracies, Manipulations & Deceptions
by Nick Redfern
ISBN: 978-1-57859-679-9

The Government UFO Files: The Conspiracy of Cover-Up
by Kevin D. Randle
ISBN: 978-1-57859-477-1

The Handy American Government Answer Book: How Washington, Politics and Elections Work
by Gina Misiroglu
ISBN: 978-1-57859-639-3

The Handy Communication Answer Book
By Lauren Sergy
ISBN: 978-1-57859-587-7

Hidden History: Ancient Aliens and the Suppressed Origins of Civilization
By Jim Willis
ISBN: 978-1-57859-710-9

The Illuminati: The Secret Society That Hijacked the World
by Jim Marrs
ISBN: 978-1-57859-619-5

The New World Order Book
by Nick Redfern
ISBN: 978-1-57859-615-7

Plagues, Pandemics and Viruses: From the Plague of Athens to Covid-19

by Heather E. Quinlan
ISBN: 978-1-57859-704-8

Secret History: Conspiracies from Ancient Aliens to the New World Order

by Nick Redfern
ISBN: 978-1-57859-479-5

Secret Societies: The Complete Guide to Histories, Rites, and Rituals

by Nick Redfern
ISBN: 978-1-57859-483-2

The UFO Dossier: 100 Years of Government Secrets, Conspiracies, and Cover-Ups

by Kevin D. Randle
ISBN: 978-1-57859-564-8

PLEASE VISIT US AT VISIBLEINKPRESS.COM

Visible Ink Press®
43311 Joy Rd., #414
Canton, MI 48187-2075

Visible Ink Press is a registered trademark of Visible Ink Press LLC.

Most Visible Ink Press books are available at special quantity discounts when purchased in bulk by corporations, organizations, or groups. Customized printings, special imprints, messages, and excerpts can be produced to meet your needs. For more information, contact Special Markets Director, Visible Ink Press, www.visibleinkpress.com, or 734-667-3211.

Managing Editor: Kevin S. Hile
Cover Design: Graphikitchen, LLC
Page Design: Cinelli Design
Typesetting: Marco Divita
Proofreaders: Larry Baker, Shoshana Hurwitz
Indexer: Larry Baker

Cover images: Shutterstock.

Paperback ISBN: 978-1-57859-709-3
Hardcover ISBN: 978-1-57859-765-9
eBook ISBN: 978-1-57859-752-9

Cataloging-in-Publication data is on file at the Library of Congress.

Printed in the United States of America.

10 9 8 7 6 5 4 3 2 1

TABLE OF CONTENTS

PHOTO SOURCES

A380Water!: p. 33.
BMC Dermatology: p. 34.
Ramana Dhara: p. 54.
Dave Hoisington/CIMMYT: p. 159.
C. Kristiansen: p. 170.
Lamid58 (Wikicommons): p. 86.
Rivers Langley: p. 156.
Library of Congress: p. 259.
LRAD Corporation: p. 60.
National Archives at College Park: p. 38.
Orange County Archives, California: p. 153.
Praxis Films: p. 30.
Shutterstock: pp. 3, 6, 10, 12, 15, 20, 22, 23, 27, 28, 37, 45, 47, 51, 63, 65,
    66, 69, 74, 77, 80, 84, 90, 94, 101, 102, 107, 109, 112, 114, 118, 122, 127,
    130, 134, 136, 142, 144, 148, 155, 161, 169, 173, 175, 176, 179, 182, 184,
    186, 188, 191, 194, 197, 199, 202, 203, 205, 206, 209, 212, 213, 216, 222,
    226, 229, 231, 235, 237, 239, 243, 247, 250, 251, 253, 255, 256, 260, 261,
    267, 269, and all chapter title and sidebar art.
Trinity Court Studio, Pittsburgh: p. 125.
U.S. Air Force: p. 61.
U.S. Congress: p. 119.
U.S. Federal Government: p. 53.
U.S. Fish and Wildlife Service: p. 138.
U.S. Geological Survey: p. 141.
U.S. National Archives, Washington, D.C.: p. 17.
U.S. National Oceanic and Atmospheric Administration: p. 99.
U.S. Navy: p. 40.
World Health Organization: p. 91.
Public Domain: p. 71.

ACKNOWLEDGMENTS

I would like to thank Roger Jänecke and his amazing staff at Visible Ink Press, especially editor extraordinaire Kevin Hile. The whole team makes me look great as a writer, and the quality of the books they publish is top-notch. I am so honored to be a part of the family.

To my literary agent, Lisa Hagan, wow! We have had so many wonderful years together as dear friends and colleagues, and you have always been the best agent a writer could dream of. Thank you!

To my family—my mom, Milly, who never fails to love and support me; my amazing sister and best friend, Angella; my talented little brother, John, who towers over me; and my dad, John, who watches from his perch in the heavens. To my good friends and colleagues; to my fans, followers, and readers; and to all the radio hosts and interviewers who have helped my career along, thank you so much!

To Pam Ramsey, my high school journalism teacher, who first told me on that fateful day that I had what it takes to be a real writer. To my pets along the way that offered their own brand of love and support even when I was writing and should have been playing with them. To those people over the years who offered me assistance with personal stories, research, and insights to use in my books. Thank you!

To the truth-tellers, including Mae Brussell, who started it all for me.

And to the one person that is my sun, my stars, and my moon all rolled into one: my son, Max. He's my everything, and everything I do is for him and his future.

INTRODUCTION:
CONSPIRACY AND TRUTH IN A TOXIC WORLD

We live in a toxic world. Whether it be the air we breathe, the clothes we wear, or the ideas and opinions of the mass media we fill our minds with, toxicity is everywhere. In our quest to make life easier, we seem to be killing ourselves to live, ingesting and injecting poisons into our bodies through the water we drink, the food we eat, and the medicines we take. We wear clothing riddled with toxic chemicals and use products that leach things into our food-stuffs we probably never intended to consume, had we known they were there.

In many cases, we know full well how we are being poisoned, and we embrace it anyway with processed foods, preservatives, water barely fit for drinking, and air that seems invisible yet teems with danger beyond our vision. It's like a bad habit we just can't seem to break.

Not a day can go by without our being exposed to a variety of toxins that wreak havoc on our physiology, our minds, and our well-being and derail our conscious attempts to achieve good health and a long life. It's gotten to the point in modern society where our toothpaste is filled with chemicals like aluminum and fluoride, both heavily linked to a number of diseases. Our clothing is treated with flame retardants that can keep us from burning up quicker in a fire but are filled with toxic substances that are getting into our lungs and causing allergies. We go outside and breathe in the pesticides just sprayed on the garden next door, never realizing the invisible damage it's causing until much, much later, when we get a cancer prognosis.

No, this book isn't intended to cripple the reader with anxiety. Yes, this book is intended to scare the living $*%t out of the reader, to get the reader to start asking questions and taking positive actions to eliminate toxins and demand our government protect us from it, rather than allow us to be poisoned for the sake of some pharmaceutical company or big agribusiness profit margin.

Everyone enjoys reading about conspiracies perpetrated by the government or secret societies, and you will find some of that in this book, but most of what you will read here is not about conspiracy theories. Oh, no. There will be a lot of facts, statistics, and scientific research and data presented here, alongside some good, old-fashioned conspiracy theory because sometimes where there is smoke, there is the presence of a former fire, or the beginning of a new one.

 There will be a lot of facts, statistics, and scientific research and data presented here, alongside some good, old-fashioned conspiracy theory....

In this case, the word is used more to describe things that are being done to humans without their direct knowledge. Things decided on and passed in laws that we often don't hear a peep about unless it's from some activist group on the internet trying to wake us out of our slumber of "my government and corporations care about me and would never hurt me" denial. Things conspired in shadows in the high halls of government, academia, and corporate conference rooms that take so much time to ever trickle down to the little people (us), if they ever do, and often when they do, people are in so much denial they refuse to acknowledge these conspirings against them.

In order to determine whether or not a conspiracy holds up to the light of truth, one must be retrained to step away from believing everything they see and hear that aligns with their own political or religious identity. People need to know how to stand back and use great discernment, enough to stop proclaiming that every word they hear on Fox News or CNN is true because both are nothing more than corporate machines that serve a political agenda. They need to learn to follow the sources to the money and to follow the money to the sources. Sadly, so few people care enough to do this. It's way too easy to just buy the party line and tow it to shore. Yet conspiracies leave clues and often turn into facts and truths. For example, there are rocket fuel compounds found in certain diabetes medications. Sound like conspiracy theory to you? Nope. As you will see later, there is ample proof.

Do you use sunscreen? Did you know that the FDA itself listed four specific ingredients found in most sunscreens that are not just toxic but are absorbed into the blood at levels high enough to pose real risks to your health? It's true. Yet many will refuse to believe it or research it on their own, deciding instead to take the easy way out and listen to everything they are told that aligns with their existing beliefs that "sunscreen good, sunshine bad." More on that later.

Conspiracies occur everywhere, in business and industry, in politics and media, even in our own lives. We are always being lied to, manipulated, harassed, persuaded, gaslighted, overwhelmed with propaganda and bias, cajoled, and even in many cases threatened (doctors shaming patients to force them to accept certain procedures, workplaces telling employees they will fire those who don't do this or that...) to keep up a particular status quo of belief. We allow the corporations and their political mouthpieces to control and govern every aspect of our lives, refusing to go against whichever political (or religious) affiliation we have signed up for, even when the evidence to the contrary is right in our faces. It's cognitive dissonance, denial, and identity/ego issues all rolled up into one big ball that makes us perfect sheeple for the wolves that want to exploit us and make money off of our hard-working backs.

Ever watch a room full of diehard Democrats tearing apart Republicans but refusing to accept the corruption of their own party? Ditto for the Republicans. Ditto for the major religious traditions of the world. Ditto for everyone and anyone who loves to point fingers at the "other guy" without first examining their own sins. That's what those who cry "conspiracy theory!" would love us to keep doing, and believing, so they can continue to pour toxins into our air, water, and food without much resistance.

We're told that we need to use sunscreen to block damaging radiation, but many sunscreens contain toxic chemicals.

Now, this author is the first to admit some conspiracy theories are way out there, but even in those cases, the power of belief is strong among those who believe them and try to convince others. We mock them and make fun of them on social media, and yet, we are all victims of actual, REAL conspiracies that have truth behind them that we continue to push the agendas of. We are all complicit until we begin to wake up, do our own research, and stop blindly trusting the authority figures that seek to keep us in the dark for their own power and profits.

This book looks at all the ways that we are being poisoned and who is behind it. The "why" is pretty easy to figure out. It's usually always money. Greed. Profits. Stockholders anxious to make more money. Corporate CEOs in bed with congressmen and women, with a lot of "you stroke my back, I'll stroke yours." The powerful and elite working overtime to protect their power and their elite status. A mass media beholden to the advertising money that keeps them in business, usually from Big Pharma, today's most powerful lobby group in the nation's political arena.

It is also a book about things that are in no way, shape, or form conspiracies. Things like the amount of plastics we are poisoning our planet with. Things like indoor air pollutants and their links to respiratory illnesses. Things like drugs and pills that have been recalled because of toxins that cause cancer or have killed innocent people. Things like what's in the food you buy that causes cancer, or why the soil most food is grown in contains toxins beyond pesticides, or how known carcinogens are found in popular breakfast cereals and so very much more. It is not fun stuff to read about, think about, or even write about, but it is necessary and important.

But we will also explore some of the more interesting conspiracy theories that may or may not be true, may be true in part, or may be total disinformation and propaganda. Yes, we will look at those, too, because this book seeks to present all sides whenever possible. It also seeks to show how the mind is poisoned by social engineering, mass media manipulation, narrative control, and the control of social media by major corporations. Freedom isn't free. In fact, we pay for it with our bodies, our minds, and our growing lack of well-being and increased rates of cancer, heart disease, autoimmune disorders, autism, and the like. We pay for it with news channels that spin our information to the bidding of their corporate backers, which we then spread on social media, calling fake anything that goes against it. We pay for it with the destructive power of identity politics and the trolling and hostility to opposing points of view. We pay for our freedom with mandated vaccines filled with toxins, air that is killing us even as we need it to live, water that is so thick with pathogens and microbes, and food so riddled with chemicals and preservatives, it's a wonder we are still standing.

 Freedom isn't free. In fact, we pay for it with our bodies, our minds, and our growing lack of well-being and increased rates of cancer, heart disease, autoimmune disorders, autism, and the like.

Most of all, we pay for freedom with censorship of any information, debate, and dialog that is not a part of the status quo or accepted by the big corporate brothers watching over us all. Did you know that Google and Amazon have links to the CIA? Were you aware that Facebook has performed social engineering experiments on its users? Why should you be aware, considering these are things done outside of the public eye—kind of like … conspiracies?

One of the biggest problems and setbacks of exposing toxicity in our culture, society, and lifestyles is the backlash. Many people who write, speak, and teach about toxins in our air, water, food, media, and medicine end up threatened, banned, blocked, removed, harassed, targeted, publicly shamed—you name it. Books like this get censored and suppressed. Whistleblowers are thrown in jail to rot while the mainstream media they attacked and exposed calls them terrorists and villains. Documentaries get banned from Amazon and Netflix. Often, social media is the last resort, other than word of mouth, for spreading information to help those find the truth, except social media is being censored heavily today by those beholden to the corporations that buy the most advertising from them, again usually Big Pharma, Wall Street, and banks.

You want the truth? The question is, can you handle the truth? Sometimes, conspiracies end up being the truth, but they first must be dragged out into the light of day and exposed so that we can see all sides and examine each angle. You will hear those who seek to fool you say that "the science is settled" or "don't confuse your Google searches with my college education" or something of that sort, not realizing that much of the damning evidence against their claims comes straight from the very corporations they are shilling for. You will hear news stories that present one side of a subject, and not another, because they can't, or they will lose millions in advertising dollars. You will see ads and stories on your social media pages picked just for you to sway you toward your political identity and away from questioning any gaps you might sense in your gut exist when the words and actions of your beloved leaders don't add up. You will be told by your chosen political party that they are your saviors when they are in bed with the enemies. You will be riddled and bombarded with lies, false flags, misinformation, disinformation, rhetoric, propaganda, and manipulation from all sides.

Where can you find the truth? This book doesn't profess to be the end-all of that question because the truth is, truth shifts as new facts and information are presented. Do you know that it was once believed sugar was good for you? Nowadays, just about every disease known to humanity is linked to inflammatory responses caused by too much sugar intake. We were once told it was safe to take certain pharmaceuticals, only to have thousands die from the side effects and consequences of "settled science." We get vaccines shot into us and our babies by the dozens today without ever asking once what is in them (you'll be shocked to find out). We buy into diets and fads and "the latest research," never bothering to identify

the origins of said research, which is often funded directly or indirectly by the very entities pushing that particular product, fad, or medical necessity.

And here is the thing: you can search for much, if not all, the information you need to make better choices from a more balanced vantage point yourself, without a medical degree or surgical license. You can outresearch many of your own doctors and professionals, who often do not keep up on the latest publications in their own fields, let alone other fields. You can google your way to knowledge of what you are injecting, ingesting, and taking into your own body and the bodies of your children, although don't use Google, as it is a propaganda machine in and of itself. It's all there, often right on the websites of the companies and government agencies themselves. Truth is often hidden in plain sight. Follow the sources to the money and the money to the sources.

Do you really know what is in that drink you're sipping or that snack you're chewing? Unfortunately, we can't trust companies to be entirely honest with us. The good news is that researching our food has never been simpler.

It's not all a big conspiracy, folks. Which is really too bad because we'd all be a lot better off if it were. We'd be a lot better off if the government bodies that are supposed to protect us weren't bedfellows with the corporations who only see us as a way to make money and who care little or nothing at all about us as human beings. We'd be a lot better off if we had companies and powerful elites who wanted to protect us, our children, and the planet instead of finding the easiest, quickest ways of keeping us sick, suppressed, and stupid with junk foods, bad air, dirty water, and chemical soups in our medications.

Chemicals and toxins that make their way into our bodies cause all kinds of harm. They dysregulate the body and alter the speed at which many functions occur. Some examples are: decreasing or increasing heart rate and blood pressure, interrupting neuron activity in the brain, decreasing or increasing hormones in the body, blocking insulin-receptor sites so sugar cannot get into the cell to produce energy, poisoning enzymes so they no longer do the work they are intended to do, displacing or destroying minerals and weakening bones, damaging vital organs, damaging and altering DNA, modifying gene expression, damaging cell membranes, and increasing toxicity in the body so that detoxing becomes next to impossible. We may not be aware of this cumulative assault on our health until it's too late, and we are suffering some major health setbacks.

But we know it's happening on a less obvious level, don't we? Look around you. Everyone is sick and tired and sick and tired of being sick and tired. Shouldn't we, with all the newest technologies and discoveries, be getting healthier and stronger? Yet, the United States is a miserable seventeenth in health, and our infant mortality rates are one of the worst in the world. We have more people in this country with cancer, heart disease, autism, Parkinson's, Alzheimer's, allergies, asthma, mental disorders, and suicides. We are one hell of a miserable country for all our supposed medical advances, access to water, genetically modified organism (GMO) farms that claim to make foods healthier, and clean air (or so we think). The hidden toxins we live with are killing us slowly in some cases, quicker in others. Yet we still stay in denial and refuse to question our government leaders, our medical doctors, our academics, the news media, and the talking heads they employ. We don't even have the courage to question ourselves when our guts tell us we shouldn't have this procedure or eat that processed food. We've been so manipulated that we don't trust ourselves anymore, and that puts us in the incredibly dangerous position of being taken advantage of by anyone who sees us as a lifelong patient or profit machine.

We've been so manipulated that we don't trust
ourselves anymore, and that puts us in the
incredibly dangerous position of being taken
advantage of....

If we all want to be able to make the best choices about what
we put into our bodies and what we put into our minds, it is im-
perative that we first learn about what others are doing to us and
take responsibility for letting them up until now. Then and only
then can we fight back and demand we alone get to choose and
that our choice must include natural medicine, homeopathy, whole
organic foods free of pesticides, clean air and water, truth in advertising, transparency in food packaging, political reps who are not
beholden to any industry, and the right to decide what is put into
our bodies and when. What you don't know can not only hurt you
but also kill you. It's time to get educated and wake up to the presence of the diverse toxin "conspiracies" and realities we all live with
on a daily basis and see how we can end, avoid, or counteract them
to preserve our health and well-being of body, mind, and spirit.

We have the inalienable right to life, liberty, and the pursuit
of happiness. Those things are not possible when we are filled with
toxins and feeling weak, sick, and hopeless. There is no life, liberty,
or happiness until we take back control of our bodies, our food and
water and air, our medical decisions, our ability to access information, and our minds and thoughts. Otherwise, we become slaves
to a system designed to make us sick and weak and keep us there.
Time to rise up and learn; at the very least, it's time to question
what you've been told, hold it up to the light, and examine all sides,
even the ones you've been afraid to look at because they will force
you to admit you may have been wrong, misled, or unaware. Remember the old saying, "there are three sides to every story: yours,
theirs, and the truth"?

You won't ever find the truth if you don't examine all sides
of an issue. It's like someone handing you a block that is blue on
all the sides you can see but purple on the opposite side you cannot.
Until you examine all sides, you won't know that one side is purple.
You might even find that the bottom of the block is green. Sometimes, the entire block will be blue. You just have to be willing to
step out of the bubble you've chosen to live in as your beliefs, per-
ceptions, and identity and turn the block.

THE AIR THAT WE BREATHE

In 1974, a band called the Hollies released a song called "The Air That I Breathe," the chorus of which expounded on how the singer only needed the air he breathed and to love a particular girl. Truth is, he could have lived without the girl, no matter how his heart felt at the time. But the air? Humans can generally survive for about three weeks without food (although recent research into long-term fasting has proven that factoid a bit shaky) and three days without drinkable water but only three minutes without air. So had our singer had a choice between the air he breathed and the girl, hopefully he would have chosen the air.

Air Pollution

Yet the invisible air we need for our survival is often filled with harmful and equally invisible toxins and pollutants that we inhale on a daily basis, whether inside our homes and businesses or outside in the "fresh air" of nature. They come in the form of solid particles, liquid droplets, or invisible gases. Don't think this is just a modern issue because our ancestors once breathed in their own brand of toxins, mainly those emanating from ongoing volcanic activity. Today, many of the invisible things we breathe in by the lungful are man-made, along with particulates of both organic and nonorganic origins. We also have biological molecules that are introduced into our planet's atmosphere via space sources.

The most pressing pollutants and toxins we deal with, which are no doubt causing a plethora of negative health effects, are ammonia, carbon monoxide, sulfur dioxide, methane, nitrous oxides, chlorofluorocarbons, pesticides, volatile organic compounds, gas emissions from motor vehicles, radioactive pollutants, and even odors from sewage plants and wastewater. Mmmm, breathe it all in. The obvious results are respiratory issues such as allergies, asthma, chronic obstructive pulmonary disease (COPD), and the like, but pollution kills and contributes to everything from heart disease to cancer. In a 2014 report by the World Health Organization (WHO), outdoor air pollution caused up to 4,000,000 premature deaths annually and, in 2012 alone, over 7,000,000 people died globally from pollution.

In recent years, indoor air pollution complications have caused illness and possible deaths. In addition, more people have flocked to urban areas, which have a higher prevalence of air pollution. However, anyone living within walking distance of a chemical, biological, or manufacturing facility, major freeways, highway systems, or near a slaughterhouse or factory farm runs the risk of breathing in more than the usual quota of toxins in the air.

China has one of the worst smog problems in the world, and this is primarily because they burn so much coal for fuel. People wear masks most of the time in cities like Beijing. While not quite as bad as China, the United States also has a serious smog problem.

Three Types

There are three types of pollutants in our air. Primary pollutants are produced by processes in nature such as volcanic ash after an eruption, gas fumes from automobile exhaust, and sulfur dioxide released from a variety of factories and manufacturing plants. Secondary pollutants are not emitted into the air via direct means. They are the result of the interactions between a primary pollutant and the air itself. Then there are those that are both primary and secondary. Human activity is, again, the culprit behind most pollutants. The leading pollutant is carbon dioxide, which is considered one of the primary greenhouse gases behind the growing problem of climate change.

Gases, chemical compounds, and particulate matter all get sent into the air, harming humans and animals, who breathe the air, and plants, trees, and flowers, which utilize the air.

Whether the source is a city full of smokestacks, factories, waste incinerators, and streets filled with vehicles, a rural coal mine, or fertilized farmland, there is bound to be environmental degradation to the surrounding land, water, and air via hazardous toxins and pollutants in varying concentrations. Even the burning of biomass, such as wood, and the use of dung and crop waste by poor and developing nations add to the pollution levels of a land mass, so we cannot just blame more urban, highly developed nations. Nature itself offers up dust from vegetation-free swaths of land, methane courtesy of cattle's digestive systems, radon gas from the radioactive decay within Earth's crust, and volatile organic compounds from vegetation exposed to extreme heat, ash, and smoke from wildfires and volcanic activity.

Air pollution was big news during the 1970s, when movies and commercials proclaimed gloom and doom from dark smoke spewing from smokestacks into the skies. In the following few decades, major concern surrounded urban smog and the ozone layer. The latter became fodder for news headlines and efforts in state and federal governments to focus on emissions and controlling the damage of a growing manufacturing and technological society, one that loved its automobiles. Everyone knew what smog was, and everyone worried about using too many fluorocarbons and making the ozone hole bigger.

What most people didn't realize, and still don't, is that there are hundreds of pollutants, with approximately 200 regulated by the Environmental Protection Agency's (EPA) Clean Air Act and other measures. Yet even in small amounts, those left unregulated can cause disease and death. This includes minute amounts of commonly found mercury, which affects the central nervous system; lead, which damages children's brains and kidneys; dioxins, which cause short-term liver, immune system, and endocrinal damage; and benzene, which is an EPA-classified carcinogen that irritates the eyes, skin, and lungs and causes blood disorders. Regulations that have been put in place may not be enough. According to the National Resources Defense Council (NRDC), stronger standards are needed to replace coal and fossil fuel–burning industries before the levels of polluting chemicals being released into the air threaten public health. They already have unless we can offset the damage with more "green" energy sources.

Smog

Smog is a mixture of smoke and fog. The term has become interchangeable with any air pollution. Coal-burning plants can create a mixture of smoke and sulfur dioxide to create a type of smog. Photochemical smog was first described in the 1950s. The worst smog occurs in larger urban areas and in basins and valleys where bad-quality air is trapped. Think Los Angeles and its San Fernando Valley, both known areas with high smog levels.

The Environmental Protection Agency (EPA) is responsible for measuring air quality indexes and reporting them to the government and public. Because smog contains ground-level ozone, sulfur dioxide, carbon monoxide, nitrogen dioxide, and other gases that can do great damage to human health, it is measured along this index to keep track of pollution levels and how much concentrated particulate matter and gases are too much. Some of the side effects of high smog levels include asthma, allergies, respiratory failure, lung diseases, and chronic bronchitis, striking children, the elderly, and the immunocompromised the hardest. Even birds, insects, and wildlife are affected by smog levels, along with trees and plant life.

Air pollution in the Los Angeles area has been a particular challenge. Not only are there a lot of cars and other sources of pollution, but the metro area is also surrounded by mountains that prevent the smog from drifting away, allowing it to build up for weeks or months unless washed away by a rare rainstorm.

Smog is worse during warmer summer months when the upper air inhibits vertical circulation, which is amplified again in basins and valleys surrounded by hills and mountains. Some of the planet's smoggiest cities include Los Angeles, New York City, Chicago, London, Toronto, Beijing, Hong Kong, Sumatra, Athens, and Mexico City. Wildfire smoke can create high smog levels in more forested and rural areas, too, creating a blanket of dirty, brown air that hangs over the environment for weeks or months.

Sadly, the United States creates a lot of its own smog and air pollution, but it also must deal with air pollution that travels from Asian nations. When combined with a warming world, that air pollution becomes more of a health threat as land/sea warming contributes to rising concentrations of aerosols in the atmosphere. China is one of the worst offenders, yet its battle to halt particulate air pollution has created more severe ozone pollution, mainly due to a pollutant called PM 2.5, which has dramatically altered the atmospheric chemistry and increased levels of harmful ground ozone pollution. Cleaning up the air is not an easy task, and it is an ongoing challenge as more people flock to urban areas, buy cars, and contribute to the rising levels of pollutants entering the atmosphere.

 Some of the planet's smoggiest cities include Los Angeles, New York City, Chicago, London, Toronto, Beijing, Hong Kong, Sumatra, Athens, and Mexico City.

Between the production of coal, the proliferation of automobiles on our streets, and the compounds placed into the air from a variety of other sources, it has been an assault on our lungs and respiratory systems. In the warmer months of spring and summer, the main type of smog and pollution comes from photochemical smog and its contribution to the ozone layer. Vehicle emissions, industrial compounds, and anything that forms a chemical reaction in the atmosphere are the hallmarks of the summer smog we often see hanging over cities like Mexico City and Los Angeles.

Winter smog is more oriented toward the use of coal, fire, and fossil fuels to heat up our homes, offices, and cars. Coal fires are one of the worst contributors, thus the push in recent decades to cut back on coal as a major fuel source in favor of natural gas and nuclear energy. Coal pollution isn't a recent phenomenon. It can be dated back to England during the Middle Ages, where it was a major source of energy and fuel. London became notorious for its thick, pea-soup smog. China is also a big coal-burning nation and has one of the worst air-quality records for its major cities, but its rural areas also use dirty coal for fuel.

When it comes to vehicle emissions, we usually think of cars, buses, and motor vehicles, but trains and planes also contribute to the dirty air we breathe in the form of airborne by-products that come from the vehicle exhaust system, spewing more carbon monoxide, nitrogen oxide, and other components of petroleum-based fuels into the atmosphere. Nitrogen oxide and other volatile organic compounds exacerbate the problem when introduced to sunlight, which triggers chemical reactions that form toxic vapors, ground-level ozone, and smog.

The type of air pollution that comes from active volcanoes is called "vog" to distinguish it from man-made smog. There are also plants that produce hydrocarbons, which give off smog that eventually affects the ozone formation. Plants in some regions emit significant amounts of volatile organic compounds that can be just as dangerous to our health as the fumes from a factory smokestack.

The Ozone Layer

Earth has a protective layer in its atmosphere that absorbs almost all of the sun's damaging ultraviolet light, preventing it from reaching Earth's surface and causing harm to living things. The ozone layer is found in the lower portion of the stratosphere from about twelve to nineteen miles above Earth. The thickness of the ozone layer changes with the seasons, and in some geographical locations, it is thicker than others. The thinning of this layer has been called the ozone hole and is the result of chlorine and bromine molecules encountering ozone molecules and destroying them.

Chlorine mainly comes from chlorofluorocarbons that are not destroyed in the lower atmosphere and instead attack ozone molecules. This results in more of a thin spot than a hole, but because it serves as a natural "sunscreen" for our planet, if it becomes too thin, all of the sun's ultraviolet B (UVB) radiation reaches Earth and increases cancer, skin diseases, and damage to the DNA of living things, including plants. UVB radiation affects the physiological and developmental processes of plant life and damages early development stages of growth in fish, shrimp, amphibians, and marine life.

The depletion of the ozone layer became a top concern in the 1970s. Ozone-depleting substances (ODS) were clearly cre-

The ozone layer is a critical part of our atmosphere that blocks harmful ultraviolet radiation from our sun. Chemicals such as chlorofluorocarbons can eat away at it, creating holes through which radiation can penetrate.

ating a thin area, prompting several countries to enforce a ban in the use of chlorofluorocarbons (CFCs) in aerosol form. Chemists Sherwood Rowland and Mario Molina discovered that CFCs from aerosol sprays could rise miles into the stratosphere where the sun's rays split them apart, allowing them to destroy the ozone molecules.

Volcanic eruptions also affected the ozone but in a much less direct and powerful way. Over the Antarctic, the ozone hole grew until more drastic measures began to repair the damage.

The 1987 Montreal Protocol, a global treaty to phase out ozone-depleting chemicals, was a turning point that raised awareness to the dangers of not taking immediate action. The Protocol succeeded in reducing chemicals and starting the healing process. Because of a reduction and bans in the use of products such as aerosol hair and deodorant sprays, the ozone layer has been getting thicker in the damaged zone over time. Thanks to some abnormal weather patterns in the upper atmosphere over Antarctica in 2019, NASA recorded the smallest ozone hole ever observed since 1982.

However, we are not out of the woods yet. We still must face the possibility of the discovery of new gases that affect the ozone layer negatively or of countries deciding to not follow the Protocol's regulations and bans to keep the layer from further deterioration. We also must take natural phenomena like volcanic eruptions and the growing threat of carbon pollution from climate change into consideration when trying to predict if the ozone threat will stay the same, get better, or get worse.

Smog Sickness

Whether it's traditional smog or pollution, we are sick because of it. Inflammation of breathing passages, decreased lung capacity, increased respiratory illnesses, and a weakened immune system are the results of our progress with technology, industry, and fuel production. Few people realize that constant exposure to pollutants can also lower birth rates, increase birth defects, and contribute to heart disease, cancer, and Alzheimer's. It's not hard to imagine how daily exposure to particulates and toxins from the air can wreak havoc on our entire bodies and create the perfect storm for chronic illness.

Pollution is measured by the size of the particulate matter. Fine particulate matter measures less than 2.5 micrograms (PM2.5) and indicates outdoor air pollution. Most pollution studies and research papers focus on this amount, but levels can reach PM10 and much higher. The Air Quality Index gives the government and public an idea of when the severity of pollution and smog can require staying indoors, wearing a mask, or avoiding exercise, but in recent times there have been few of those extreme precautions required here in the United States. Other countries such as China and Mexico aren't always so lucky.

A 2017 study by the American Thoracic Society found that air pollution reduces sleep quality. Global studies link air pollution to depression and, at higher levels, suicide.

Pollution doesn't just do damage to the body. It also damages the mind and emotional state. A 2017 study by the American Thoracic Society found that air pollution reduces sleep quality. Global studies link air pollution to depression and, at higher levels, suicide. The meta-analysis was published in the December 2019 issue of *Environmental Health Perspectives*. One of the study authors, Isobel Braithwaite from the University College of London, wrote that the particulate matter could be causing "substantial harm to our mental health as well, making the case for cleaning up the air we breathe more urgent." The results consistently showed that long-term pollution exposure at the PM2.5 level increased the risk of depression for suicide in children by 19 percent and that once in the body, these smaller-sized particulates could be found in the brain and other organ systems. They also caused a small increase in the risk of suicide. PM10 levels found to be too large are absorbed into the bloodstream and affect the brain.

Despite the passage of the Clean Air Act in 1970, pollution in America continues to harm millions of people due to pollutants exceeding federal standards. The lack of enforcement of the subsequent attempts at regulating pollution producers such as industrial sites, factories, and even factory farms has succeeded only incrementally, with emphasis on finding replacements for coal, but fall short when it comes to overall protection from toxins in the air, mainly due to strong lobbying from the industries involved to keep the status quo and keep regulations soft, which is often completely ineffective.

Most weather apps on cell phones now include a pollution or smog index and warnings if levels are above normal. This is especially

President Richard M. Nixon signed the Clean Air Act into law on December 31, 1970. The legislation reduced levels of lead, carbon monoxide, ozone, nitrogen dioxide, sulfur dioxide, and particulates by 70% over the next 50 years.

important for children, pregnant women, the elderly, those with immune system illnesses or compromised immune systems, and those already suffering from respiratory diseases. The truth is, pollution harms all of us, no matter how healthy we are. The WHO offers stringent guidelines for exposure to damaging pollutants, but these are not enforced, thus their 2014 estimate that every year, premature death caused by air pollution hovers around the 7,000,000 people mark, and new studies published in March 2019 raised that number to 8,800,000 people dying from lung cancer, heart disease, immune disorders, respiratory illnesses, and others linked to pollution and toxins.

Here is a list of diseases and illnesses associated with smog and outdoor pollution:

▲ Depression and suicide

▲ Asthma

▲ Allergies

▲ Sinus infections

▲ Rhinitis

▲ Inflammation of the airways

▲ COPD

▲ Emphysema

▲ Cancers of the lungs, mouth, throat, and esophagus

▲ Heart disease and chest pains

▲ Eye dryness and infections

▲ Lung and pulmonary diseases

▲ Low birth weights in infants

▲ Preterm births

▲ Higher mortality rates in infants

▲ Short-term memory loss

▲ Learning disabilities

▲ Cognitive impairment

▲ Immune system impairment

▲ Kidney disease from lead exposure

▲ IQ impairment in children

▲ Blood disorders

▲ Damage to the liver

▲ Damage to the endocrine system

▲ Attention deficit hyperactivity disorder (ADHD)

Indoor Air Pollution

So much for getting outside in the "fresh air"! Indoor air pollution, though, is just as damaging to our health, maybe even more so because of all the time we spend indoors nowadays. Whether we are working in an office or cubicle, a factory or warehouse, or we are kids in school classrooms or sitting at home playing video games, there are plenty of toxins we are being exposed daily. These toxins contribute to poor health in many of the same ways outdoor pollution does, but often, being inside exposes us to more chemicals from cleaning agents, laundry detergents, air fresheners, paint, flooring, fumes, mold, dust mites, and a host of other things.

One of the biggest types of dangerous, and often deadly, indoor pollution comes from cooking with open fires and old stoves that use kerosene, wood, crop mass, animal waste, and coal. This occurs more in rural areas and among the poor, and it contributes to close to 4,000,000 premature deaths, according to the WHO. It also contributes to diseases such as cancer, heart disease, stroke, ischemic heart disease, low birth weight, tuberculosis, laryngeal cancer, COPD, and lung cancer, and the sad fact is that close to half of pneumonia deaths among children happen because of inhaled household particulate matter.

Household fuel combustion and indoor air quality from household cooking and heating fuels occur because of a lack of electricity and modernization, as well as a lack of proper ventilation, which affects over one billion people. Lack of ventilation alone concentrates pollutants into a small space, where people tend to spend most of their time. Older homes also have lead paint and volatile organic compounds (VOCs) being released from flooring and carpeting.

No matter where you work or live, these VOCs are everywhere. New buildings contain high levels of particulate matter being released from brand-new flooring, paint, carpeting, wood, mattresses, bedding, sealants, plastics, glues, and other materials used in construction. Carpeting and plywood both emit formaldehyde gas. Paint and solvents emit VOCs as they dry. Lead paint becomes a dust that can be inhaled. The release of these chemicals into the room or building is called "off-gassing."

Then there is mold, dust mites, and carbon monoxide from faulty vents, blocked chimneys, and using a charcoal grill indoors or in areas of minimal ventilation. Asbestos, a deadly particulate, is still being found in older homes, and the abatement process is an extensive and expensive one, for those who can afford it. Pets give off dander and create havoc for those with allergies and skin sensitivities. Hair sprays, spray perfumes, aerosol cleaners, laundry detergents,

Looks like a clean, safe environment, right? Your living room is likely full of pollutants from sources such as paint fumes, formaldehyde (in everything from furniture to laminate flooring), cleaning agents, mold, mildew, dust mites, and even air freshener.

cleansing agents, and a host of other toiletries contribute to the bad air quality in the home. In the office, there are copier and ink fumes, hazardous materials, mold, and other pollutants. In the warehouse or factory, there are all types of dangerous pollutants and particulates being inhaled, often accompanied by shoddy ventilation systems.

 Indoor air is worse than outdoor air by the sheer fact that we spend more time indoors than outdoors.

Indoor air is worse than outdoor air by the sheer fact that we spend more time indoors than outdoors. Yet say the word *pollution,* and few people will mention how sick they feel at the office or how their allergies are awful at home. They continue to think of smokestacks and brown smog above a big city.

While the WHO and EPA work on outdoor pollution levels, it's up to the individual, family, or business to focus on indoor pollution. Let's not forget the schools we send our children to, which are rampant with indoor pollutants and, in older areas, lack of good ventilation systems to combat it. Again, in poorer areas the problem is exacerbated, but even richer areas tend to forget about the air they are breathing unless they can see the smog or smoke. Pollen is easy to target as a natural outdoor pollutant because you can see it, but indoors, you cannot see the dust mites in your bedding that harm your respiratory functions. You can see mold, but you cannot see carbon monoxide.

These are things many people can change, but for the poor, the lack of money or lack of access to electrical plants and better, cleaner sources of fuel make it difficult. Learning about the types of pollutants we are exposed to indoors and outdoors can help us avoid unsafe situations and places. We can stay indoors when the smog index is high or wear a mask if it is extreme. We can cut back on driving our cars and use public transportation or ride-sharing more. We can properly dispose of hazardous home wastes such as cleaners and toxic agents instead of putting them in the trash to go to the landfill. We can stop using sprays and spritzers and use balms, lotions, and products that don't end up in the air for us and others to inhale.

We can buy products that are allergy-free, keep our chimneys clean, and get our ventilation systems checked once a year. We can install carbon monoxide detectors, plant more trees, and buy houseplants known to help detoxify the air. We can vacuum and dust more to keep our kids' bedrooms free of dust mites and dander. We can buy electric cars or hybrids to cut back on fossil fuel consumption. We can buy meats and dairy products from small farms and not smog-producing factory farms.

Obviously, we cannot live lives free of all pollution, but we can certainly bring our exposure down and improve our health by doing so. When the Clean Air Act or other environmental acts and laws are introduced in Congress that aim to protect our health by improving the quality of the air we breathe, we can fight for their passage and not stay silent as major industries and corporations lobby for their own purposes. We can demand that our most precious and necessary commodity, the air that we breathe, stays breathable.

Greenhouse Gas, Climate Change, and Pollution

Ozone, carbon dioxide, and methane are all greenhouse gases. These are gases that trap heat in the atmosphere and lead to warmer temperatures, more extreme precipitation, more extreme weather, higher pollen activity, more mold, rising sea levels, and higher rates of infectious disease transmission, including the appearance of diseases like malaria in places it had never been before. Climate change adds to pollution problems because increased carbon levels are one of the main drivers of climate change today and one of the reasons why the ozone layer is not completely protected from future damage. We dodged a bullet before with the ozone hole, avoiding what could have been millions more deaths from cancer and a host of agricultural disasters, thanks to strict regulations. But a lot of people in and out of our government hate regulations and believe companies and corporations should be permitted to freely pollute our air for higher profit margins. A lot more people are too lazy, sick, or tired (probably because of the health effects of pollution exposure) to make the changes necessary in their own lives, like carpooling, walking, or riding a bike to work instead of driving alone, that would help rather than add to the problem. We are indeed creatures of habit, and those habits may be killing us.

Carbon monoxide detectors are an important way to monitor one of the most dangerous air pollutants that could invade your home environment.

"While we've made progress over the last forty-plus years improving air quality in the U.S., thanks to the

Clean Air Act, climate change will make it harder in the future to meet pollution standards, which are designed to protect health," says Kim Knowlton, senior scientist and deputy director of the NRDC Science Center, quoted in "Air Pollution: Everything You Need to Know" by Jillian Mackenzie in November 2016. The more our population grows on this planet, the harder it will become to keep that progress in check, especially with greater numbers of people living in urban areas and driving cars, traveling by plane, eating more beef, and creating more waste products.

Knowlton and her colleagues point out that greenhouse gases must continuously be kept in check, including methane from natural and industrial sources, carbon dioxide from fossil fuels, and other such gases. We emit more carbon dioxide into the air thanks to an increase of cars and vehicle emissions, but methane is far more potent and destructive. Think about this. The human population is exploding, and that means more people to feed, more people on the road driving cars, more people using heating and cooking fuels to save money, more people living in poor areas and urban centers, more land being turned into cattle grazing for more meat, more cattle emitting their own methane, and more factory farms to feed more people. More pollutants to expel into the air and in turn be inhaled.

Still, the Montreal Protocol showed the world that we can diminish a huge threat, as it did with the growing ozone hole, and proved we can succeed at reducing carbon pollution and other particulates that harm us and our planet.

We have no idea what a changing climate will ultimately mean for our air quality or the new challenges we will face with regulations and control, but it now behooves us all to do what we can starting in our own homes and neighborhoods to step it up for clean air.

Tire Pollution?

Tire particle waste is another form of pollution most people don't think about. Tires must be replaced often, so what happens to the particles when tires wear down on the roadways? A study by the European Commission looked at the negative impact of tire particle waste on air quality and

Burning rubber, one can easily see tire particles coming off a tire. But even with normal usage, tiny bits of tire are being worn off daily, and you likely breathe some of them in.

determined that 10 to 30 percent of tire material is lost as they wear down, and some of that waste is in the form of particles small enough to be inhaled.

Tire pollution is more of a problem in urban and high-traffic areas, but it can be mitigated by keeping windows closed and using the recycled-air button, using cars less, and living in areas that are not close to highways or busy roads. You should also keep tires properly inflated to avoid extra wear and tear.

Remember, three minutes without air, and we are dead. Clean air is not an option or a luxury. It's a necessity for our survival.

CHEMTRAILS AND GEOENGINEERING

Look up in the air! It's plane contrails! Or is it plane chemtrails? What is the difference, and is it possible chemicals being sprayed from drums attached to the bottoms of planes are responsible for the increased rates of allergies, asthma, chronic obstructive pulmonary disease (COPD), and other respiratory issues? Who is making those weird crisscross lines across the skies? Are they doing it for good, as in geoengineering the weather? Are they doing it for evil, as in conspiracies of population culling via poisoning the air we breathe? Are they pesticide spraying, and are we the pests?

And who are *they*, anyway? Corporations that want us sick so they can force their healing pills and vaccines on us? Government agents who want to get rid of populations so there is more good stuff for them? Sinister military groups that are a part of an even more sinister agenda to create a zombielike workforce that won't raise their voices and fight back? All of the above?

There are those who insist planes leave behind white trails in the sky that are harmless. Contrails are nothing more than a white, visible vapor trail or "condensation" trail that is the result of airplane engine exhaust. These contrails are made up of water vapor or tiny ice crystals and remain in the sky for several seconds to a few hours, depending on the weather conditions at the time. Certainly, plane contrails are common, easy to spot, and usually represent much of what people see when they look up.

Much, but not all. More and more, people are reporting high-altitude "chemical" trails that make long and elaborate lines in patterns across the sky that fail to dissipate after the normal evaporation time of contrails and, in some cases, drop down to the ground as a fine, white powder. The author of this book experienced this in the year 2001 while standing outside holding a baby. A plane crisscrossed overhead, and a fine, white powder fell to the ground and onto our revealed skin. It also left a thin veil of white powder on the ground, like patches of newly fallen snow, which quickly dissipated or dissolved. And no, this author doesn't live where it snows. Within a day, both this author and her baby were sick with terrible respiratory issues and discovered other neighbors had gotten sick as well from the mysterious powder.

 Chemtrails can include pesticide spraying and, at times, overspraying into residential areas, which happens more often than people think and has resulted in several lawsuits.

Sound like simple contrails to you? Chemtrails can include pesticide spraying and, at times, overspraying into residential areas, which happens more often than people think and has resulted in several lawsuits. It can include crop dusting, cloud seeding, and even firefighting chemical drops that are carried on the wind. Chemtrails also occur at altitudes and in weather conditions that make them impossible to be called contrails; for instance, hundreds of witnesses have observed chemtrails that shut off midflight, something contrails simply cannot do unless the engines themselves are shut down. You also have to ask why the skies are blanketing in these bizarre lines and patterns every single day since planes do fly every day? Instead, as this author witnessed, they occur either sporadically or, in the author's case, every three or four days, after which, strangely, the usual sunny, dry weather would become drizzly. Even rainy. Should we just discount all these witness reports as delusions, as many in the media and scientific community tell you to do?

Chemtrails are real, and thousands of people have photographed them, gotten sick from them, and documented them all over the world. Yet this admission doesn't mean a company or the government is out to drop some killer virus on your neighborhood. It doesn't mean they are part of a huge plot to depopulate the planet, as some conspiracy theories suggest. However, it's easy to start going down that rabbit hole the more you research the subject

Contrails are omnipresent in our skies, and they even spoil scenic views like this one. Are they harmless, or could they be chemtrails that are making people sick?

matter away from the mainstream "they don't exist" websites and news sources, not to mention that our government and military have engaged in dumping toxic chemicals into our air before. Jim Marrs writes in *Population Control: How Corporate Owners Are Killing Us* of the almost 300 secret experiments conducted between 1952 and 1969 alone on targets such as Fort Wayne, Indiana; San Francisco; St. Louis; Oceanside, California; and Minnesota's Chippewa National Forest.

The U.S. military performed what Marrs said it called "vulnerability tests" using chemical and biological compounds in open-air testing on civilians since the 1940s. Chemtrails are not new, and there is plenty of fodder to feed the fires of conspiracy. Marrs quotes the late Dr. Ilya Sandra Perlingieri, a college professor and environmental activist, as saying, "For more than a decade, first the United States and then Canada's citizens have been subjected to a 24/7/365-day aerosol assault over our heads made of a toxic brew of poisonous heavy metals, chemicals, and other dangerous ingredients. None of this was reported by any mainstream media." She goes on to say that the U.S. Department of Defense, along with the military, have systematically blanketed American skies with chemtrails, also known as "stratospheric aerosol geoengineering," which are not the usual plain contrails.

Yet government agents, politicians, corporation executives, and military officials have repeatedly gone on record to state that chemtrails are not real. The mass media, which is beholden to the above, parrots the cry of "conspiracy theory," thereby nullifying and humiliating the reports of those who suggest otherwise. Try googling chemtrails, and see what you get. You will have to search high and low to find an article that doesn't automatically dismiss them as nothing but conspiracy theory unless you use a search engine that isn't operated by the National Security Agency (NSA) and Central Intelligence Agency (CIA), that is.

In 2001, U.S. representative Dennis Kucinich (D-OH) was running for president. On the campaign trail, he was asked if chemtrails were real. He flatly stated, yes. A few years prior, as chairman of a subcommittee of the House Committee on Oversight and Government Reform, Kucinich introduced the Space Preservation Act (HR 2977). It was created to ban exotic space weapons, but it included in its language a ban on any chemical, biological, climate-changing, or tectonic weapons, including chemtrails. Why he would have introduced legislation that included something that did not exist—and asked to ban, to boot—is beyond logic.

Congressman Dennis Kucinich (D-OH) introduced an ultimately unsuccessful bill in 2001 to ban exotic space weapons, which included anything using "electromagnetic, psychotronic, sonic, laser, or other energies" as well as anything "expelling chemical or biological agents" directed at people.

One way the media dismisses the subject is with one-sided reporting of research studies. One such case came from *New Scientist*, which featured "Chemtrails Conspiracy Theory Gets Put to the Ultimate Test" by Phil Plait in its August 2016 issue. The article documents a study done by seventy-seven atmospheric scientists, including some researchers from the University of California, Irvine, and the Carnegie Institute, who surveyed hundreds of experts on contrails and atmospheric deposition by presenting them with evidence from chemtrail websites, including videos, articles, photos, and water and soil analyses. The consensus was that all but one scientist/researcher believed these were anything other than contrails and natural water. The one dissenting scientist reported high levels of atmospheric barium in a remote area with standard "low soil barium." Barium, by the way, is one of the chemicals associated with geoengineering (more on that later).

But first, the writer dismissed all who believe in chemtrails, including those of us who have experienced them and their effects, and labeled them conspiracy nuts. However, there is a fatal flaw in the kind of research that relies on evidence found on a website. One must wonder if the results of these scientists would have been the same had they stood under a crisscross sky and experienced the strange, white mist or taken immediate water and air samples.

In 1996, the U.S. Air Force (USAF) published a report about weather modification, and soon, rumors abounded of the USAF spraying chemicals in the air to change the weather and offset the coming threat of climate change. This forced a response from both the Environmental Protection Agency (EPA) and the National Aeronautics and Space Administration (NASA), who tried to offset rumors and conspiracy theories with reports and media responses to the contrary. The USAF claimed the whole conspiracy theory started because of its fictional scenario paper called "Weather as a Force Multiplier: Owning the Weather in 2025," which it presented as a method for teaching and strategizing future scenarios. The CIA, NASA, and the National Oceanic and Atmospheric Administration (NOAA) came together in 2013 to provide funds to the National Academy of Sciences to study geoengineering as a method to counteract climate change and adamantly stated "chemtrails" would not be part of that method. Yet they admitted to researching geoengineering, which includes "cloud seeding"—this could be interpreted as the chemical alteration of cloud formation to induce rain, could it not? So, were they just admitting there were no "chemtrails" at the same time they were admitting the method of cloud seeding chemicals existed? It's all semantics, no?

Edward Snowden, the former CIA whistleblower, even admitted on the *Joe Rogan Experience* podcast that he never found anything in his database searches about chemtrails.

One must wonder whether he would have come to a different conclusion had he searched for "chemicals found in plane contrails used in geoengineering."

Famous whistleblower Edward Snowden warned American citizens about how the government was spying on them, but even he said he found no evidence about chemtrails being used.

In an effort to provide fairness and balance, let's take a look at the arguments for both sides. We know that NASA, NOAA, the NSA, the Department of Defense (DOD), the CIA, the Department of Energy (DOE), and the military insists that chemtrails are nothing more than misidentified contrails (just like UFOs are misidentified military prototypes). We know the majority of scientists interviewed by mainstream media think chemtrails are nothing but a conspiracy theory perpetuated by those with no scientific background. We know that the Google searches turn up dozens of articles that deny chemtrails exist and shame anyone who questions that conclusion.

Atmospheric scientists have established that plane contrails can stay in the air for hours, even dissipate into that fine mist or veil that those who report chemtrails speak of. The varying sizes of ice crystals in the contrails descend at different rates, spreading the contrails out vertically, and varying wind speeds and other weather factors, as well as altitude, come into play to create the horizontal spreading. This creates what looks to those on the ground like cirrus uncinus clouds. With enough air traffic in the sky at one time, it can also create that blanket of mist or overcast skies that lasts for hours.

Because these are all things reported by chemtrail witnesses, these scientists then conclude that everyone is seeing normal contrail activity. They point to fact sheets produced by NASA, the EPA, NOAA, the Federal Aviation Administration (FAA), and the USAF that show the scientific nature of contrails and analyses made since 1953 and disprove the existence of any chemicals. These fact sheets confirm that chemtrails are a hoax. In fact, the same Wikipedia site makes the generalization that all samples taken by chemtrail believers are usually flawed because they don't automatically rely on government or academic laboratories. The idea here is that normal people don't know the proper scientific method for gathering air, water, and soil samples or know where to send them to be properly analyzed.

Analysis of commercial airline tracks makes the case that they are not fit for chemtrail spreading or climate engineering (yet no mention of noncommercial aircraft, but we will let that go for now). If you are inclined to get your news and truth off of Wikipedia, that's where this information came from. There are sources, of course. All of them are from science magazines, journals, and websites that come—we must be transparent—from a pre-existing "chemtrails are a conspiracy theory" stance. Normally, the pro-science and government/military stories are going to come from pro-science and government/military sources. Some come from people who claim anyone who believes in chemtrails is an idiot and should be punched in the face, or worse.

Yet there is still smoke, and where there is smoke, we must look a little closer. Not every scientist agrees, and not every test shows there are no chemicals being sprayed in our skies.

In 2007, KSLA-TV news in Shreveport, Louisiana, did an investigative report where they took water samples captured during a chemtrail "crosshatch" and had them tested in a lab. The result was a high level of barium, which is an alkaline earth metal rarely found in nature because it oxidizes in the air. Yet the test samples showed abnormally high levels, and state health officials confirmed the toxic dangers of that level of barium on the human nervous and immune systems. But no one was willing to go on record and blame the chemtrails. In 2002, environmental engineer Therese Aigner found large amounts of barium, aluminum, calcium, magnesium, and titanium in three rainwater and snow samples. Aigner also went on record to say the tests showed a "very controlled delivery (dispersion)" by chemtrails in the area the samples were taken from and that this violated more than a half dozen state and national laws and regulations.

 If there are higher rates of aluminum being found in places the public is reporting chemtrails, shouldn't someone try to take that seriously?

Studies done in the Mt. Shasta, California, region show high levels of the same chemical elements found in Shreveport, mainly aluminum, which for years prompted a public demand for state and local officials to at least take the time to review both sides of the scientific literature. The dangers of aluminum will be presented throughout this book, but suffice to say it is linked to everything from autism to dementia to neurological diseases to amyotrophic lateral sclerosis (ALS, more popularly known as Lou Gehrig's disease) to Alzheimer's to nervous system damage, loss of memory, and a host of other illnesses. If there are higher rates of aluminum being found in places the public is reporting chemtrails, shouldn't someone try to take that seriously? Instead, locals have been met with accusations of being dumb, uneducated morons who believe in conspiracies. Yet the fact that someone or something is responsible for the unusually higher levels of barium and aluminum is not a conspiracy at all. It is a truth no one yet is being forthcoming about. No wonder the public has so much mistrust for science, government, and the military.

The scientific and medical communities, as well as our politicians, consider it pure coincidence that so many people are getting sicker with ailments linked to higher concentrations of toxins and chemicals in our environment. Not one official explanation has yet truly explained the increased respiratory issues experienced by those exposed to these "fake chemtrails."

Chemtrail Flu and Morgellons

The pharmaceutical industry has been accused many times of making people sick so they can sell them more drugs. It's become a popular meme theme on social media, poking fun at the largest lobbying group in politics for purposefully drugging and vaccinating people to make them sicker than ever so they can then sell more prescription drugs to take care of all those sicknesses. It sounds like conspiracy theory, but more people are suspect of Big Pharma than ever, thanks in part to billions of dollars being paid out in various lawsuits because of toxic side effects, including deaths, due to prescription drugs, vaccines, and unnecessary or botched medical procedures or medical errors, which are the third leading cause of death. More on that later.

But does this same suspicion—that the government, aided by Big Pharma, wants people sick so it can then make more money off them—have a connection with chemtrails? According to some outspoken doctors, yes. The term *chemtrail flu* refers to the many illnesses people report after being exposed to chemtrails. A former university medical researcher named Dr. Leonard Horowitz now wears a target on his back because of his outspoken claims that chemtrails are the result of the government conspiring with pharmaceutical companies to sicken large segments of the population and turn them into customers for life. He believes they do this by spraying chemical concoctions into the air and that since the 1990s, a strange, flulike illness has been on the rise.

On his website, Horowitz says, "We have seen this type of epidemic since the end of 1998 and the beginning of 1999. People

Photos like this one showing an Airbus A380 loaded with barrels have been presented as evidence of the chemtrail conspiracy. In reality, these barrels are used as balast to simulate passenger weights during flight tests.

have been hacking and coughing with this bizarre illness that does not seem to follow any logical virus or bacterial onset and transition period." He goes on to mention additional symptoms, such as sinus congestion, sinus drainage, malaise, and fatigue but no fever. He believes "chemtrails are responsible for a chemical intoxication of the public, which would then cause a general immune suppression, low grade to high grade, depending on exposure."

Horowitz studied the content of chemtrails sprayed over his Idaho home and claims the Centers for Disease Control and Prevention (CDC) confirmed they contained ethylene dibromide, a carcinogen once removed from unleaded gasoline because of health concerns. So why was it added back to these high-altitude military craft that are spraying it? He believes it's to make us all sick.

Obviously, Horowitz was attacked from all sides for his views and opinions, including the idea that vaccines could cause sterility in some populations, which, by the way, has been proven true. (See the vaccine section later in this book.) When such claims seem so outlandish to our normal sense of what we believe happens in our world, it's easy to write someone off as a quack or conspiracy theorist, yet are his claims really that outlandish? Would military craft ever do such a thing on behalf of our government officials in order to boost pharmaceutical sales or keep people sick and tired and weak, so they don't fight the system? It sounds right out of a science fiction novel. The question is, is it entirely fictional? The common chemicals associated with chemtrail samples contain nanosized compounds that cause intense reactions and inflammation in the body. These nanoparticles even cross the blood–brain barrier, bringing aluminum into the brain and possibly causing neurodegenerative diseases like ALS, Alzheimer's and Parkinson's, all of which are on the rise in the United States.

One such affliction often linked to chemtrails is the mysterious skin ailment known as Morgellons. According to Jim Marrs in his book *Population Control: How Corporate Owners Are Killing Us*, one chemtrail ingredient is said to be tiny, synthetic filaments called polymers. These polymers can cause serious skin infections and, according to polymer chemist Dr. R. Michael Castle, who has studied polymers related to chemtrails, may be the cause of the malady known as Morgellons, which infects as many as 12,000 American families.

Morgellons has no explanation according to modern science, yet it produces disfiguring sores, lesions, and an awful sensation of something moving or crawling just beneath the skin. Some victims

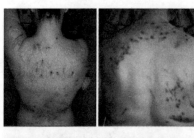

find strange fibers coming out of their skin, which match materials reportedly from chemtrails. Yet the government denies Morgellons exists, and those who suffer from it are told they have a mental illness. Doctors tell victims they may be suffering from delusional parasitosis, which means they believe they are infected with something and manifest it by compulsively picking and scratching at their skin. Of course, that doesn't explain the fibers.

Morgellons is believed to have first occurred in the 1600s, according to written accounts that sound similar to the modern reports. Dr. Randy Wymore, an associate professor of pharmacology and physiology at Oklahoma State University, has claimed to have extracted the fibers that grow under the skin, which he states do not match any known artificial fibers. The vast major-

The top photos here show lesions on the backs of Morgellons patients; the bottom photo is a close-up of the fibers found embedded in a patient's skin.

ity of fibers reported are black, red, or blue. Are these, as many conspiracy theorists claim, nanoparticles or synthetic model organisms being sprayed down on humanity in the form of chemtrails?

One of the most famous victims of Morgellons was the legendary singer/songwriter Joni Mitchell, who told the *L.A. Times* in 2010 about the strange condition she suffered from. She also wrote about it in her 2014 memoir. Her symptoms got so bad, she couldn't wear clothing. She told the *Times* that she had fibers in a variety of colors protruding out of her skin "like mushrooms after a rainstorm." Mitchell devoted much of her time after that trying to advocate for victims of Morgellons. The disease got a lot of mainstream media airtime in 2006, as thousands of others came forward to talk about their own bizarre symptoms. Again, some victims were told they were delusional and needed mental or psychiatric evaluation. In 2007, a victim of Morgellons started the Charles E. Holman Morgellons Disease Foundation to raise awareness and provide grants for research.

The name *Morgellons* comes from a seventeenth-century term that describes the same type of illness that was once recognized by the medical community. Whether or not it is being perpetuated today by chemtrails, it has a basis in medical history, minus the weird fibers now being reported.

Nowadays, only those who suffer continue to push for it to be taken seriously, as the medical community in general calls it a disease of delusion. In 2012, the CDC studied the illness and found no basis for its existence. It referred to Morgellons as a "delusional infestation." Tell that to the victims who continue to report the same mysterious symptoms and the strange presence of fibers that can't be explained by sheer delusion.

To date, conspiracy theorists continue to demand that the government come clean about what those fibers are and if there is a connection between nanotechnology and chemtrails. The government and medical community continue to insist that chemtrails, nanotech fibers, Morgellons, and a secret plot to use chemicals dropped from planes to debilitate the populace and sterilize the poor and unworthy do not exist and never have existed.

To date, conspiracy theorists continue to demand that the government come clean about what those fibers are and if there is a connection between nanotechnology and chemtrails.

Officially, we are told that there is no solid proof that there is a connection between these afflictions and illnesses and chemtrails. We are also told that chemtrails don't exist, and we are simply mis-identifying contrails. But there is plenty of circumstantial evidence, and more doctors and researchers are asking about the correlations between the lines in the sky that don't follow the rules of contrails and the ever-climbing rates of illnesses that are linked to whatever chemicals are making their way to the ground and onto our bodies.

What *Are* They Spraying? And Why?

Richmond County Daily Journal contributing columnist Rob-ert Lee, a Marine veteran and concerned citizen, wrote in March 2018 about chemtrails. His column "Chemtrails: What Are They Spraying?" was his own personal experiences researching the sub-ject and his frustrations with the powers-that-be insisting chemtrails were contrails, and all who saw them were simply mistaken. Lee captures the tone of the debate when he writes, "I can't tell you what our government is truly doing but I can say I do not believe it is for the benefit of our people. I can tell you what I have found out. Because of the persistence of the chemtrail conspiracy theory and questions about government involvement, scientists and gov-ernment agencies around the world have told their citizens that the chemtrails are in fact contrails.... B.S." He goes on to explain his own conversion from someone who didn't care about contrails to someone knowing the difference between contrails and what he was seeing and someone who knew the planes that were spraying were not airlines but rather military.

Lee remarks about the recent admissions of those working in the geoengineering field that chemtrails are real and involved in weather modification. But he has more questions, as do we all. Why are we being sprayed? Is it about fighting climate change? Or is there a more sinister purpose for dropping chemicals on human be-ings en masse?

Instead of name-calling, ridiculing, and shaming those who believe chemtrails are real, perhaps we should be asking ourselves if the government and military we trust so blindly have ever sprayed us with toxic chemicals from planes in the past and tried to cover it up. The answer is yes, they have.

Government-Approved Aerial Spraying of U.S. Citizens

At the height of the Cold War in the mid-1950s, the U.S. Army used motorized blowers positioned on top of a low-income housing building in St. Louis, Missouri; at various schools; and from the backs of station wagons to spray a dangerous chemical compound into the air in predominantly black neighborhoods. These tests were repeated ten years later, all as part of a government biological weapons program that became public in 1994. The Army chose the poor neighborhoods because they most resembled the Russian urban areas the U.S. government was prepared to attack if needed.

The chemical compound consisted of zinc cadmium sulfide in the form of a fine powder. The Army may also have combined those particles with radioactive particles, according to the research of a Southern Illinois University at Edwardsville sociology professor, Lisa Martino-Taylor, although she offered no proof of this.

Martino-Taylor had received documentation of the spraying via the Freedom of Information Act, revealing that three-quarters of the people sprayed were African American. Years later, people were dying of cancer at high rates and at relatively young ages. The Army did admit they sprayed the area from tops of buildings, but some residents reported green Army planes flying overhead and dropping a powdery substance onto the people, including children, below.

When the secret testing came to light in 1994, it spawned a committee study by the National Research Council, which determined three years later that no one was harmed by the chemicals. Yet higher levels of cadmium and long exposure periods are known to cause lung cancer and damage kidneys and bones. No follow-up studies by the Army appear to have ever been conducted.

On September 20, 1950, the people of San Francisco became unsuspecting victims of chemical spraying when a U.S. Navy ship off the coast used a giant hose to spray a microbe cloud into the fog. The military was testing the effects of a biological warfare attack on a major city. The spraying took place over the course of seven days, during which time one person died and countless others fell sick. This was one of the first large-scale biowarfare tests on the public, which overall lasted for twenty years.

Of course, barrels of zinc cadmium sulfide would never be labeled so obviously, but this chemical, combined with radioactive materials, may have been sprayed over civilians by the U.S. Army, according to one researcher.

During those twenty years, 239 tests would be conducted over various cities, many documented in *Clouds of Secrecy: The Army's Germ Warfare Tests over Populated Areas*, written by Dr. Leonard Cole, the director of the Program on Terror Medicine and Security at the Rutgers New Jersey Medical School. One of the two types of bacterial agents the Navy ship sprayed over San Francisco was *Serratia marcescens*, which had never been seen in local hospitals. A patient named Edward Nevin died from an infection of the bacteria in his heart.

Information was revealed in the 1970s of the decades of horrific experiments of people at the hands of the military and government. Biological and chemical weapons were sprayed throughout the nation in the form of bacterial agents and chemical powders from low-flying airplanes that would begin their journey near the Canadian border and fly south over American cities to dump their toxins.

Cole wrote in his book that the military sprayed a school and tried to excuse it by claiming they were creating a way to mask a city to protect it from terrorists. Sadly, the powder they did this with included toxic chemicals such as cadmium. Evidence showed the powder was evident days later for up to 1,200 miles from the initial drop zone.

In 1966, subway riders in New York City were exposed to light bulbs filled with bacterial agents as part of a test called "A Study of the Vulnerability of Subway Passengers in New York City to Covert Attack with Biological Agents." The light bulbs would break, releasing the bacteria on the tracks to see how they dispersed, engulfing people as the trains pulled away. The National Academy of Sciences later analyzed the experiments and claimed the bacterial agent used, *B. globigii*, was a pathogen and a major cause of food poisoning. It is also sometimes fatal.

In 1966, light bulbs full of the bacteria B. globigii, which causes food poisoning, were deliberately shattered in New York City subways in an immoral experiment on unwilling and unwitting civilians.

In 1977, military officials testified before Congress and claimed these experiments were necessary to show vulnerabilities in dealing with biological warfare. It didn't matter if the public consented or not because the military claimed it was for the greater good of research. These experiments are just a handful of the many that occurred throughout our history, especially during the Cold War era, and served as a historical precedent for the "chemtrails" argument.

▲ The United States held aerial biological and chemical weapons tests in at least four states (Florida, Alaska, Hawaii, and Maryland) during the 1960s to develop defenses against such weapons, according to declassified Pentagon documents.

▲ From 1963 to 1969, the U.S. Army sprayed several U.S. ships with bacterial and chemical warfare agents as part of Project Shipboard Hazard and Defense (SHAD). Ship personnel were neither notified nor equipped with protective clothing during the tests. They were exposed to nerve gas VX, sarin, sulfur dioxide, and cadmium sulfide, among other agents.

▲ In 1949, the U.S. Atomic Energy Commission released iodine-131 and xenon-133, both radioactive agents, into the atmosphere near the Hanford Nuclear site in Washington state, contaminating three small towns as part of Operation Green Run.

▲ U.S. military crop dusters engaged in spraying deadly sarin gas and VX over the Wiltshire countryside in the United Kingdom during the late 1960s.

▲ Thousands of U.S. military personnel were subjected to mustard gas from 1942 to 1944 to study the effectiveness of gas masks and protective clothing.

▲ In 1957, nuclear explosions in the atmosphere above Nevada released radiation that caused up to 212,000 cases of thyroid cancer and up to 21,000 subsequent deaths. During the early years of the Cold War, the United States, United Kingdom, and Australia engaged in atmospheric nuclear tests to determine how much fallout would make Earth uninhabitable. They collected and studied the bones of the human bodies across the globe, mainly infants, to see how they absorbed the contaminants. This was an ultra-top-secret program because of the fact that parents and family members were never told why their dead relatives' bodies were being taken.

▲ The USS *George East-man* was used as an observation monitor point for nuclear and biological tests during the 1950s and 1960s. Pentagon documents revealed over 3,000 military personnel were exposed to toxic chemical and biological agents during the 1960s.

The USS George Eastman, *a Liberty class cargo ship, is shown here in 1956 after being refitted with scientific equipment designed for nuclear tests.*

▲ The military admits there are over four different types of chemtrails, according to a popular conspiracy site, Rense.com, which also claimed air traffic controllers were concerned about chemtrails.

▲ Today, conspiracy theorists point to chemtrails, mass forced vaccinations across the globe, toxins in food and water, the release of new and deadly pathogens, 5G and directed-energy weapons, and genetically modified organisms as the new methods of testing toxins on human subjects with or without their consent or knowledge.

Rules, Codes, and Ethics

Our history is filled with torture, experimentation, and cruelty in the name of scientific research and medical progress. After World War II, Nazi experimenters and torturers were put on trial for war crimes. These trials were overseen by the United States under President Harry Truman, the United Kingdom, and the Soviet Union. They took place in Nuremberg, Germany, in November 1945 and became known as the Nuremberg Trials, which led to a code of ethics, specifically ten points making up a code by which "permissible medical experimentation" could take place.

 During human experiments of the past, doctors and medical personnel performed the heinous acts on subjects, including children, the disabled, pregnant women, and prisoners, without asking themselves if they were doing harm.

The Nuremberg Code was originally ignored until about twenty years later and is now considered a global code of ethics and morals that heavily influences clinical research, scientific study, and human rights. Other such codes of ethics include the Declaration of Helsinki, originally signed in Helsinki, Finland, in 1974 as a code of medical research ethics involving the use of human experimentation. It has been revised several times. The Hippocratic Oath is an oath of ethics taken by physicians. It originated somewhere around 275 C.E. in Greece and is still referred to today as a set of moral rules to be followed by those practicing medicine. The most important tenet of the oath states "first do no harm," yet as we will see in following chapters, there are many in the medical industry who seem to have forgotten this in their quest for profits and money. During human experiments of the past, doctors and medical personnel performed the heinous acts on subjects, including children, the disabled, pregnant women, and prisoners, without asking themselves if they were doing harm. They treated their subjects like objects and cared nothing for their health, well-being, or lives, harkening back to the experiments by Nazis on their prisoners, only this was going on in America.

The Declaration of Geneva, another medical ethics charter, was adopted in 1948 by the General Assembly of the World Medical Association and revised numerous times, as late as 2017. Its purpose was to affirm guidelines for human rights and patient rights as recognized by those in the medical community.

The Hippocratic Oath

I swear by Apollo Physician, by Asclepius, by Hygieia, by Panacea, and by all the gods and goddesses, making them my witnesses, that I will carry out, according to my ability and judgment, this oath and this indenture.

To hold my teacher in this art equal to my own parents; to make him partner in my livelihood; when he is in need of money to share mine with him; to consider his family as my own brothers, and to teach them this art, if they want to learn it, without fee or indenture; to impart precept, oral instruction, and all other instruction to my own sons, the sons of my teacher, and to indentured pupils who have taken the physician's oath, but to nobody else.

I will use treatment to help the sick according to my ability and judgment, but never with a view to injury

The Nuremberg Code

1. The voluntary consent of the human subject is absolutely essential. This means that the person involved should have legal capacity to give consent; should be so situated as to be able to exercise free power of choice, without the intervention of any element of force, fraud, deceit, duress, over-reaching, or other ulterior form of constraint or coercion; and should have sufficient knowledge and comprehension of the elements of the subject matter involved, as to enable him to make an understanding and enlightened decision. This latter element requires that, before the acceptance of an affirmative decision by the experimental subject, there should be made known to him the nature, duration, and purpose of the experiment; the method and means by which it is to be conducted; all inconveniences and hazards reasonably to be expected; and the effects upon his health or person, which may possibly come from his participation in the experiment. The duty and responsibility for ascertaining the quality of the consent rests upon each individual who initiates, directs or engages in the experiment. It is a personal duty and responsibility which may not be delegated to another with impunity.

2. The experiment should be such as to yield fruitful results for the good of society, unprocurable by other methods or means of study, and not random and unnecessary in nature.

3. The experiment should be so designed and based on the results of animal experimentation and a knowledge of the natural history of the disease or other problem under study, that the anticipated results will justify the performance of the experiment.

4. The experiment should be so conducted as to avoid all unnecessary physical and mental suffering and injury.

5. No experiment should be conducted, where there is an a priori reason to believe that death or disabling injury will occur; except, perhaps, in those experiments where the experimental physicians also serve as subjects.

6. The degree of risk to be taken should never exceed that determined by the humanitarian importance of the problem to be solved by the experiment.

7. Proper preparations should be made, and adequate facilities provided, to protect the experimental subject against even remote possibilities of injury, disability, or death.

8. The experiment should be conducted only by scientifically qualified persons. The highest degree of skill and care should be required through all stages of the experiment of those who conduct or engage in the experiment.

9. During the course of the experiment, the human subject should be at liberty to bring the experiment to an end, if he has reached the physical or mental state, where continuation of the experiment seemed to him to be impossible.

10. During the course of the experiment, the scientist in charge must be prepared to terminate the experiment at any stage, if he has probable cause to believe, in the exercise of the good faith, superior skill and careful judgement required of him, that a continuation of the experiment is likely to result in injury, disability, or death to the experimental subject.

Why anyone would be shocked by this is surprising, considering the U.S. government has for decades engaged in experimentation and testing on human beings, using such methods as biological, chemical, nuclear, and psychological warfare; poison; starvation; torture; water deprivation; sleep deprivation; mind control; and more. The government, often working in conjunction with the military or universities and members of academia, has experimented on orphans, pregnant women, prisoners, the disabled, mentally ill patients, and just about anyone who cannot fight back, most often without their consent. Subjects have been force-fed poisons, including pesticides and known toxins under the guise of "safety research." But aside from the terrors our leaders and scientists have engaged in behind the closed doors of laboratories and testing facilities, the spraying of toxins happens in plain sight, sometimes even with public approval.

and wrongdoing. Neither will I administer a poison to anybody when asked to do so, nor will I suggest such a course. Similarly, I will not give to a woman a pessary to cause abortion. But I will keep pure and holy both my life and my art. I will not use the knife, not even, verily, on sufferers from stone, but I will give place to such as are craftsmen therein.

Into whatsoever houses I enter, I will enter to help the sick, and I will abstain from all intentional wrong-doing and harm, especially from abusing the bodies of man or woman, bond or free. And whatsoever I shall see or hear in the course of my profession, as well as outside my profession in my intercourse with men, if it be what should not be published abroad, I will never divulge, holding such things to be holy secrets.

Now if I carry out this oath, and break it not, may I gain for ever reputation among all men for my life and for my art; but if I break it and forswear myself, may the opposite befall me. (Translation by W. H. S. Jones.)

(NOTE: The phrase "First do no harm" was believed to be a seventeenth-century translation of the phrase "I will abstain from all intentional wrongdoing and harm.")

There are millions of doctors, scientists, and researchers who take these oaths and codes of ethics to heart, but there are many who do not, forgetting that they are there to serve the patient and the public. Punishment for breaking the Hippocratic Oath among doctors and medical personnel is not always enforced, except when the courts step in during malpractice suits. As for the more powerful entities that engage in human experimentation, we know from history they don't always play by the rules, or they change the rules to fit their needs as they arise.

Aerial spraying of insecticides and pesticides is common, especially in agricultural areas around the country that grow fruit and vegetables. This spraying is done to protect against mosquitoes and the many viruses they carry, such as Zika, which was the main culprit of spraying in the recent past. This is not a conspiracy. State government websites must post about aerial spraying to let the public know, and possibly avoid, being sprayed as a result. The CDC has information on aerial spraying. Their Zika virus page states at the top, "When people in a large area are getting sick or when large numbers of mosquitoes are found, airplanes can be used to treat

A worker sprays plants to kill mosquitoes that might carry the Zika virus. Such programs are commonly announced to residents before they are implemented.

very large areas with insecticides safely, quickly, and effectively. This process is called aerial spraying." These spray information pages describe the pesticides being used, how to cover plants and fishponds, and how to avoid contact with your skin.

According to these fact sheets, the EPA has claimed all the pesticides used are safe despite suggesting you stay indoors and avoid contact or breathing in the chemicals. None of the fact sheets references a pesticide that is unsafe, and yet, as we will learn in a later chapter, pesticides are carcinogenic and toxic. So, if a spraying is announced for your neighborhood, it behooves you to stay indoors and not touch the stuff. If you come into contact with it, either via the skin or inhaling, go to the ER.

In February 2018, a judge ordered the California Department of Food and Agriculture (CDFA) to stop using aerial pesticides in a statewide program until the department complied with environmental laws. Eleven citizen groups came together in the city of Berkeley to sue the state after many years of trying to put an end to the spraying of seventy-nine different pesticides, many known to cause birth defects and cancer and to kill off bees, butterflies, fish, and birds. There was never any public notice of the decisions by the state to use these poisonous chemicals. Before the injunction stopping the spraying, the state's attorney admitted that the CDFA had already carried out over 1,000 pesticide spraying programs! The judge not only ruled that the CDFA had no right to spray toxic chemicals on people, but it had no right to do so without informing them.

The chemicals being sprayed included chlorpyrifos, which causes brain damage in children and endangers wildlife; methyl bromide, a toxic fumigant that depletes the ozone layer; and a chemical warfare agent called chloropicrin, which causes genetic damage. Sprayed pesticide often drifts, so even if it's targeted over specific areas where there are no homes, it has the potential to make its way to residential areas, especially if it's windy out.

So, the next time the EPA or CDC tells you they are spraying your area for this mosquito or that pest, and that the chemical agents they plan to use are safe, don't believe them. It happened in California for years, and it happened in every other state, and will continue to do so until challenged by concerned citizens demanding the truth about what is being sprayed from planes and that it be stopped for good.

In January 2020, the European Food Safety Authority banned sales of chlorpyrifos as years of research indicated that there's no "safe" level of exposure. In the United States, some individual states—including California and Hawaii—have banned its use on farm lands. Several other states—including Massachusetts, New York, Vermont, and Oregon—have sued the EPA for not enacting a nationwide ban. But, as you might suspect, The Dow Chemical Company lobbying efforts have "persuaded" the EPA that the toxic and potentially deadly chemical could still be used in U.S. food production. This is a terrible example of corporate greed over the need for public safety.

—NaturalHealth365.com

Conspiracies abound about the use of chemical and biological agents being sprayed in the skies to cull the population or keep people sick and weak so that pharmaceutical companies can make money from lifelong patients by selling them the pills they need to get better from the ailments caused by the spraying toxins. Chemtrails are often linked to these spooky programs, which will be covered in another section. Regardless of whether those conspiracies hold any truth, there are laws and codes restricting the use of testing on human subjects with chemical and biological agents, such as U.S. Code 1520a, which prohibits testing or experimenting involving such agents on human subjects. However, the code includes these exceptions to this rule:

1. Any peaceful purpose that is related to a medical, therapeutic, pharmaceutical, agricultural, industrial, or research activity.

2. Any purpose that is directly related to protection against toxic chemicals or biological weapons or agents.

3. Any law-enforcement purpose, including any purpose related to riot control.

As you can see, these exceptions are broad and pretty much nullify any restrictions the code might offer, which should strike terror into the hearts of every citizen in this country. Part C of the code states that the secretary of defense can conduct such tests and experiments only if informed consent to the testing is obtained from every human subject in advance. We know from past experiments on humans that this does not happen and that the code rules have been violated on several occasions. We also know that the verbiage of this section implies that the government can use the fact that it informed the public in advance as the actual "informed consent" and conduct the experiments even if the subjects protest or refuse. Some question if the very fact that these laws and codes are publicly available to be read by anyone in advance (search for U.S. codes or go to https:// www.law .cor nell.edu/us-code.text/50/1520a), that qualifies as informed consent via a type of acquiescence or submitting or complying silently without protest.

The most chilling part of Code 1520a is the reference to biological agents, including natural and man-made viruses, bacteria, fungi, pathogens, infectious substances, micro-organisms, and synthesized components of said organisms. This means they exist and can

Water and soil tests where chemtrails have been noticed reveal unusually high levels of barium and aluminum.

be weaponized, if they haven't already, whether in the form of chemtrails or something else.

Despite ongoing denial that chemtrails exist, numerous books, documentaries, and YouTube videos show people claiming otherwise as well as dangerously high barium and aluminum levels found in water and soil samples from areas where chemtrails were seen. This would indicate the public is being experimented on without its consent by being forced to ingest and inhale such particulates, which are linked to a number of ailments and diseases such as memory loss, lack of coordination, dizziness, nausea, impaired thinking, tremors, headaches, flulike symptoms, lethargy, and muscle aches, all associated with just one ingredient—aluminum. Or are those who fly these planes just completely oblivious to what's coming out of their engines and falling to the ground below?

According to the EPA Air Quality Management Rule 402: Nuisance—"A person shall not discharge from any source whatsoever such quantities of air contaminants or other material which cause injury, detriment, nuisance, or annoyance to any considerable number of persons or to the public, or which endanger the comfort, repose, health or safety of any such persons or the public, or which cause, or have a natural tendency to cause, injury or damage to business or property." So, when you look up and see lines of white across the sky, dissipating slowly into a thin veil of haze, perhaps you might want to go indoors and wait until whatever is being sprayed is gone. Or you can take your chances and assume it's contrails and breathe deeply from the fresh air. Your call.

WEATHER WARFARE

The debate over chemtrails almost always leads into a larger debate about weather modification and geoengineering. Geoengineering is the fancy word for climate intervention using specific technologies to intervene and offset some of the adverse effects of climate change. Though most climate scientists agree that trying to influence weather patterns will help lessen the effects, they mainly agree that it won't stop climate change, mainly because there is no way to fully predict the consequences and side effects to the environment and to the health of the living things on the planet.

Conspiracy theorists believe geoengineering is being conducted via the chemical mixes pouring into the skies in the form of chemtrails. Many theorists also believe the goal is not so much to help adjust weather patterns to produce more rain where needed or lessen the punch of extreme weather patterns but to systematically poison the populace as part of a population control plan to eliminate millions, even billions, of people to preserve dwindling natural resources. As wild as that sounds, there is some reason to wonder. But first, more on climate engineering.

There are two methods by which the climate might be manipulated to better serve humanity and offset the damages of decades of pollution, greenhouse gases, carbon dioxide, and methane being released into the atmosphere. In most places on the planet, weather patterns are changing, and more parts of the globe are experiencing extreme precipitation conditions or extreme drought. Climate change was mistakenly called "global warming" but is more about the existing weather patterns in a region shifting or becoming more extreme, amplifying existing conditions that were once normal and changing regions from temperate to arid, from arid to tropical, and so on. It's not just about warming temperatures during the day but also warmer temperatures at night when the earth is supposed to be cooling. To most scientists, the main "driver" of climate change is human activity, and when combined with natural cycles, we are told we are facing a threat of unprecedented proportions.

This is why geoengineering is exploring two different techniques in the deliberate large-scale intervention of Earth's own natural systems. One is solar radiation management (SRM). This technique aims to reflect a small proportion of the sun's energy back out into space to counteract rising temperatures from greenhouse gases. These gases absorb energy in the atmosphere and trap it there, raising temperatures. Some of the ways SRM works:

▲ Space reflectors: These block a small part of the sunlight before it can reach Earth.

▲ Albedo enhancement: This technique increases reflectivity of clouds and the surface of Earth to reflect more of the sun's energy back into space.

▲ Stratospheric aerosols: Small, reflective particulate matter is put into the upper atmosphere to reflect sunlight before it reaches Earth. This is the big thorn in the side of chemtrail conspiracy theorists who believe these aerosols contain dangerous chemicals that make their way back to the surface of Earth.

▲ Greenhouse gas removal (GGR) or carbon engineering: This technique sets out to remove excess greenhouse gases, mainly carbon dioxide, from the atmosphere to counter increased levels of greenhouse gases and their effects on ocean acidification.

GGR techniques include:

▲ Afforestation: A global effort to plant more trees and offset rampant deforestation.

▲ Ambient air capture: Large machines capable of removing carbon dioxide from ambient air.

The difference between afforestation and reforestation is that afforestation refers to planting trees where none have stood for a considerable time, whereas reforestation (pictured) is planting trees where they were recently lost by harvesting or fire.

▲ Ocean fertilization: Adding extra nutrients to the ocean to draw down excess carbon dioxide from the atmosphere.

▲ Enhanced weathering: Exposing minerals to react with the carbon dioxide in the atmosphere and divert the resulting compound back into the ocean and ground soil.

These and other techniques can only work if all nations of the planet were to enact them because local and national efforts alone would not be enough.

Many conspiracy theorists don't believe in climate change, although it is quite obviously happening. They argue over whether humans are the cause, but one thing is certain. Geoengineering would not exist if climate change didn't exist. Clearly, the governments of the world feel certain that if they don't intervene, the planet is in big trouble. Whether climate changes are anthropocentrically driven, or natural, or both, it doesn't matter. Must we resort to tampering with weather patterns and conditions we don't fully understand to try to fix the problem? According to chemist and former British government chief scientific advisor David King, who was quoted at the Clean Growth Innovation Summit in 2018, "Time is no longer on our side. What we do over the next 10 years will determine the future of humanity for the next 10,000 years." We are running out of time to mitigate the damage and may be well beyond the tipping point to stop it, these scientists tell us. People ask if geoengineering the weather is just a way to buy more time for a species that refuses to change their habits and caused the damage in the first place.

 We are running out of time to mitigate the damage and may be well beyond the tipping point to stop it....

To emphasize the general distrust the public has in the government and in science, we can point to a 2017 group study by the Cooperative Institute for Research in Environmental Sciences (CIRES). The study included members of the public from a variety of nations, including the United States, Japan, Sweden, and New Zealand, who were asked their opinions about the various techniques involved with geoengineering, such as carbon sequestration and cloud brightening. The participants then responded with questions and concerns, such as "What happens if these technologies backfire with unintended consequences?" "Aren't these just treating the symptoms of climate change but not the cause?" "Shouldn't we change our lifestyles and patterns of consumption and make geoengineering the last resort?" Such intelligent questions showed that the public is more aware and knowledgeable on the topic than we thought, and they are not entirely trusting of the official solutions, at least not without more information and clarification.

These participants were more in favor of mitigation and adaptation over altering our weather with techniques we don't know the consequences of. The more extreme beliefs involved in climate change conspiracies take that distrust one step further and ask whether our government and those of other nations are trying to affect weather on a larger scale for their own purposes? Whether or not you believe in climate change, it is changing. The specific drivers of that change are still being researched, examined, and debated, but those drivers may not be as critical to conspiracy theorists as the "why" behind them. Why is climate change happening? Why is geoengineering occurring in our skies? Is there something else going on, such as the militarization of the weather for warfare purposes? Let's go down that rabbit hole.

Weather Warfare and HAARP

Weaponizing our weather is not new. Experiments date back to the 1950s, according to Jim Marrs in *Population Control*, and Dr.

Edward Teller, the father of the hydrogen bomb, had once proposed seeding the upper atmosphere with millions of tons of sulfur or other heavy metals to create cloud cover that would deflect the sun's rays from reaching Earth's surface. But Teller admitted it would be a hard sell to the public because of the potential to pollute the air with dangerous and toxic particulate matter.

In 1996, the Air Force published *Weather as a Force Multiplier: Owning the Weather in 2025*, in which it determined it could undertake an offensive program to modify the weather and warned against letting any enemy nation do the same. In 2013, the Council on Foreign Relations released a publication expounding the fact that nations had not made much progress dealing with emissions and that a geoengineering program was not out of the question. The council even discussed the use of solar radiation management (SRM).

The concept of injecting heavy metal particles into the air from chemtrails is often linked to the High-Frequency Active Auroral Research Program (HAARP), which the Air Force claims was shut down in 2014. The Gakona, Alaska, main facility, with alleged satellite facilities in Colorado, Texas, California, Germany, Norway, Wales, Japan, Russia, and Australia, according to some conspiracy theorists, has been the fodder of conspiracy theories for decades, claiming it is involved in triggering extreme weather conditions, earthquakes, blackouts, communications systems failures, and more.

HAARP's original intention was to research the ionosphere's properties and behavior by exciting a limited area of the ionosphere with high-frequency transmissions using the Ionospheric Research Instrument (IRI) in a controlled manner. Instruments at the HAARP observatory would study the data to get a better understanding of the ionosphere's behavior under solar stimulation. HAARP also examined the potential for developing new ionospheric-enhancement technology for radio communications and surveillance as part of a joint project with the U.S. Navy, Air Force, University of Alaska at Fairbanks, and the Defense Advanced Research Projects Agency (DARPA).

It was HAARP's association with the shadowy DARPA that most likely began the conspiracy theories of deeper reasons for the facility's existence and the kind of work conducted there. Because HAARP used a high-powered,

A view of the High-frequency Active Auroral Research Program (HAARP) site in Mount Sanford, Alaska. The program was implemented to study Earth's ionosphere.

high-frequency transmitter that could influence the ionosphere, and much of its research was classified from public view, conspiracies abounded of more sinister purposes, including the pulsing of frequency signals to make people sick, weather warfare and manipulation, generating earthquakes and tsunamis, even mind control and behavior modification via electromagnetic manipulation of the ionosphere.

In "The Ultimate Weapon of Mass Destruction: Owning the Weather for Military Use," Professor Michel Chossudovsky wrote in July 2018 that "environmental modification techniques" had been used for decades. He claims the research began with American mathematician John von Neumann working with the Department of Defense in the late 1940s, and they foresaw a coming time of climate-based warfare. In 1967, during the Vietnam War, Project Popeye utilized cloud-seeding techniques to create a longer monsoon season and block enemy supply routes on the Ho Chi Minh Trail.

The technology to alter major climate patterns and affect weather anywhere in the world comes from initial research done at the HAARP facility as an appendage to the Strategic Defense Initiative (SDI), aka Star Wars. Prof. Chossudovsky continues: "Weather-modification, according to the U.S. Air Force document AF 2025 Final Report, 'offers the war fighter a wide range of possible options to defeat or coerce an adversary.'" He claims this includes the ability to create floods, hurricanes, droughts, even earthquakes!

A Catholic nun, environmental scientist, and epidemiologist, Dr. Rosalie Bertell warned that a system as powerful as HAARP could disrupt the ionosphere and subject the planet to dangerous radiation.

Conspiracy theorists believe these same things today could be caused by HAARP despite claims it is officially shut down. In 1977, the UN Framework Convention on Climate Change was ratified, banning "military or other hostile use of environmental modification techniques having widespread, long-lasting or severe effects." This included "any technique for changing—through the deliberate manipulation of natural processes—the dynamics, composition or structure of Earth, including its biota, lithosphere, hydrosphere, and atmosphere, or of outer space." However, many researchers point out that discussion of the subject is considered taboo, especially when referring to activities today that might be considered a part of the manipulation of weather, climate, and tectonic activity. One might ask, if it's been banned, why is no one allowed to discuss it at an official level? And those who do bring up possibilities that this is still taking place today are labeled wild and crazy conspiracy theorists?

One of the first to question this was Dr. Rosalie Bertell, a world-renowned scientist who served on the Bhopal and Chernobyl medical commissions, who said, "The ability of HAARP/Spacelab/rocket combination to deliver very large amounts of energy, comparable to a nuclear bomb, anywhere on earth via laser and particle beams, are frightening. The project is likely to be 'sold' to the public as a space shield against incoming weapons, or, for the more gullible, a device for repairing the ozone layer." Bertell, who published "Background on the HAARP Project," likened HAARP to a gigantic heater that can cause "major disruptions in the ionosphere, creating not just holes, but long incisions in the protective layer that keeps deadly radiation from bombarding the planet." In the year 2000, Bertell told *The Times* of London that the U.S. military was working on weather systems as a potential weapon, including enhancing storms and diverting vapor rivers in the atmosphere to trigger severe flooding and droughts. To even think that these goals would be cut off in 2014, when HAARP's main facility was shut down, is to believe the government and military would throw away millions of dollars, perhaps hundreds of millions, of research and potential weapons technologies.

Could HAARP have been responsible for horrific droughts in Cuba, Afghanistan, Syria, Iraq, and Iran, all nations interestingly at odds with U.S. agendas? According to the Tyndall Centre for Climate Change Research at the University of East Anglia, a 2017 study showed that Asian countries from Kazakhstan to Saudi Arabia would experience warming twice as much as other nations and that these areas, including Iran, Afghanistan, and Tajikistan, would face major famine. In 1999, Iran suffered the worst drought in over a century. Not only did this weaken the nation's economy, already devastated by sanctions and bombing by U.S. allies, but it made the area incredibly vulnerable to further manipulation. The same thing was occurring in Syria and Iran.

Coincidence? Natural causes? Or possibly the effects of a manipulated effort to alter weather conditions for the worse? As for the toxic chemtrails issue, in March 2013, the U.S. Patent and Trademark Office granted a patent to Rolls-Royce plc for developing a method of preventing contrails from forming using an electromagnetic wave generator. This would also prevent the formation of artificial clouds. Other patents have been granted in the past to cover up contrails. Why, if they are nontoxic and safe? Clearly, there has been ample interest in masking chemicals sprayed from planes altogether.

But why? According to Leonard Horowitz, author of *Healing Codes for the Biological Apocalypse* and *Emerging Viruses*, who began to investigate chemtrails after witnessing and photographing spray planes over his northern Idaho home, the agenda is to make a lot of people sick. In his books, he documents the links between chemtrails and eugenics, or genocidal depopulation, and claims big banking and royal families—the "elite"—are behind the plan to cull the public with sprayed chemicals and vaccines. In William Thomas's *Chemtrails Confirmed*, he writes that Horowitz believes the agenda behind chemtrails "is to completely deplete and weaken entire populations so that people are not too sick to work, but too wasted to do much else—like protest chemtrails, fraudulent presidential 'selections,' illegal and immoral wars, and the hijacking of the Constitution." And he is not alone in this line of thinking, as we will examine in an upcoming chapter on Agenda 2030.

 The use of electromagnetic manipulation against a population to render them sick, weak, and ineffective at fighting back is even more desirable because it eliminates the need for planes....

Thomas also writes about the links between the study of eugenics, the investigation of genetic differences in populations and genetic predispositions of diseases in those populations, and the Rockefeller Foundation's eugenics program in the 1920s that focused on mass sterilization and depopulation of minorities in America.

The use of electromagnetic manipulation against a population to render them sick, weak, and ineffective at fighting back is even more desirable because it eliminates the need for planes and witnesses to the visible trails in the sky and avoids having to explain away the samples of white powder on soil and in water. HAARP and its ilk are the ultimate weapons of mass destruction, except for vaccines, as we will explore later. HAARP cannot be seen, or tracked, or proven as the reason why we as a species are getting sicker, dying younger, and failing to reproduce fast enough.

Even if HAARP was indeed shut down in 2014, there is no doubt in the minds of conspiracy theorists that other facilities exist all over the globe with the same or greater potential to turn energy into weapons and alter human minds, thoughts, behaviors, and actions from a remote location. It even has the power to render people ill with a host of diseases like cancer, heart disease, autoimmune disorders, chronic fatigue, and respiratory issues.

The toxins may not be chemical but rather energetic, as in microwaves, particle beams, directed energy, plasma, lasers, and heat rays—the stuff of science fiction now firmly entrenched as science fact, courtesy of our governments and militaries in their quest for dominance of the planet, the people, and the atmosphere.

DIRECTED-ENERGY WEAPONS
AND ELECTROMAGNETIC WARFARE

The invisible threat of directed energy is a clear and present danger that many people insist is responsible for a host of negative effects on individuals and populations. Directed-energy weapons are the new frontier of warfare, fought with lasers, microwaves, particle beams, sound frequencies, heat, plasma, and other highly focused energy sources. The Pentagon, the Defense Advanced Research Projects Agency (DARPA), and various government and military research laboratories and facilities, including the High-Frequency Active Auroral Research Program (HAARP) facility and its satellites, are all engaged in the proliferation of directed-energy weapons and applications.

Electromagnetic and energy warfare is not toxic in the same sense as a chemical pollutant, but the effects on the human body and mind are equally poisonous. Some believe energy manipulation is more dangerous because it works at the most fundamental of levels in human physiology to change, alter, or damage genetic components. On the battlefield and in riot control situations, directed-energy weapons, sometimes in the form of nonlethal weaponry, have the singular goal of disabling or debilitating human beings, if not killing them outright.

Sound and heat have been utilized by law-enforcement agencies during riot control. For instance, the Long Range Acoustic Device (LRAD) is employed to cease violent behavior by temporarily disabling victims with sound waves. In December 2006, a

A Long Range Acoustic Device (LRAD) is seen here mounted on the USS Donald Cook.

Freedom of Information Act request revealed a government study using microwave auditory effects to disrupt behavior in people without them being aware of what was happening. Microwave hearing would sound like voices in someone's head. This technology, called Voice to Skull (V2K), was researched as far back as the 1950s and patented in the 1970s. The U.S. Navy conducted research in 2003 with a system called Mob Excess Deterrent Using Silent Audio (MEDUSA), which involved pulsing microwave signals into the minds of people to disorient and debilitate them.

Microwave beams are the notorious directed-energy weapon of choice for many people who call themselves targeted individuals (TIs). These people claim to experience a number of symptoms associated with directed-energy weapons, such as:

▲ Muscle vibrations and tingling

▲ Humming in the head

▲ Voices in the head

▲ Clicking sounds inside the skull

▲ Dizziness

▲ Heat rashes on skin

▲ Sinus issues, allergies, and throat soreness

▲ Confusion and vertigo

▲ Out-of-body sensations

▲ Debilitating autoimmune diseases and cancer

TIs often are whistleblowers who feel they are being electronically harassed and stalked to shut them up or keep them from going public with damaging information about a corporation or politician, but often, these individuals have no idea why they are being targeted. They claim they are guinea pigs for the testing of directed-energy-weapons. They also claim you can shield yourself against directed-energy weapons by wearing leather or other protective clothing, including rubber shoes/boots to offset the electrical sensations or protect against microwaves.

There is precedence for their claims. The Russian LIDA machine in use in the 1980s utilized electromagnetic pulses that, when directed at humans, made them weak, exhausted, and disoriented. Machines like the LIDA

Sometimes known as a "heat gun," the Active Denial System is a vehicle-mounted device that is often used to disperse crowds by heating up surfaces, including human skin.

can also pulse heat, sound, and light toward the target. The Magnetically Accelerated Ring to Achieve Ultra-High Directed Energy and Radiation (MARAUDER) is a plasma weapon that accelerates a torrid of plasma at a significant percentage of light speed toward its target. Who can forget Ronald Reagan's Strategic Defense Initiative (SDI) program in the 1980s, also known as "Star Wars," which sought to use space-based lasers to destroy enemy targets?

The Active Denial System (ADS) is a directed-energy weapon using lasers. The Air Force developed it and used it in the Iraq War. Built by Raytheon Corporation, the ADS produced millimeter wavelength bursts of energy that could penetrate 1/64th of an inch into the human skin, just enough to agitate water molecules in the dermis and produce a powerful heat sensation that could stop a person in their tracks.

California Burning

Directed-energy weapons were blamed by conspiracy theorists and some firefighters for the rash of northern California wildfires in 2017 and 2018. The media and officials ignored the anomalies associated with the fires, but many small, independent researchers reported them, enough to start the smoke of a conspiracy fire. Even firefighters commented on how "hot" these fires burned, three times hotter than the average house fire, and how aluminum and glass melted, which normally doesn't happen in a typical wildfire. Yet plastic trash cans were not scathed at all.

Aerial drone images pointed to a burn line so straight it appeared to have been cut with a laser into the ground. Additionally, there were no entry points for some of the fires, no marks or charring on streets the fire moved across (even with houses on either side burned to the ground), and YouTube has videos of trees that appeared to have burned from the inside out. Some reported seeing animals burned to death mid-run, as if zapped by a laser rather than burned by fire, in which case they would have eventually fallen to the ground and died. In Paradise, California, locals claimed the fire made a clean perimeter around the town. They also pointed to the fact that officials blamed the fires on climate change and dry, combustible brush, yet many trees were green and lush. They also pointed out inconsistencies in official reports that the vegetation was dry at times and wet at other times and that ceramic tubs and sinks with a higher burning point (from 1,800 to 3,000 degrees Fahrenheit) than wood (374 to 500 degrees Fahrenheit) had been obliterated to dust by the flames, yet the trees remained.

Photos and videos appeared of alleged beams of light or high-energy lasers (HELs) hitting the area at night. This has led conspiracy theorists and concerned citizens alike to question whether microwave-beamed energy weapons caused the fires. Fires that plagued Northern California over the last few years have been linked to the behavior of directed microwave radiation, or microwave amplification by stimulated emission of radiation (masers). Retired California firefighter John Lord and other investigators reporting independently of mainstream media claimed they saw guardrails catch fire where metal bolts connected to the wood, one of the hallmarks of a directed-energy weapon. Others posted pictures of cars and houses burned, but the vegetation amid them was untouched.

Accusations were even made that smart meters contributed to fires, whether intentionally or mistakenly.

The "why" is even more immersed in conspiracies. Why would anyone want to start horrific fires that destroy land and kill people and wildlife? Theories range from:

▲ Testing out a directed-energy weapon (DEW) in more public settings

▲ A mistakenly aimed DEW during a military test

▲ A terrorist attack

▲ The desire to create a narrative from the event that benefits certain corporations or the government (there is one theory that suggests the area was desired for a high-speed rail system and eminent domain was not an option)

▲ The desire to cover up or destroy evidence of something else entirely.

China Lake Quakes

In 2020, over 8,000 fires burned 4,000,000 acres, destroying homes and killing dozens in California.

Though the last one seems incredibly far-fetched, some people point to the recent earthquakes (2019) near China Lake in California, which allegedly served to destroy underground child-trafficking bases and possibly chemtrail toxin holding facilities. The earthquakes were claimed to have been engineered by a HAARP DEW to appear as authentic seismic activity, much like that associated with fracking. People pointed to the ability of our own military to use a laser weapon from a satellite or plane to create a powerful, pinpointed earthquake.

The China Lake Naval Base was struck by two strong quakes in July 2019, which became known as the Ridgecrest Quake, and subsequent aftershocks. The first quake struck on July 4, measuring 6.4 on the Richter scale. Most imagined that was the main quake, but the next day, a second quake hit, measuring 7.1. The damage done to the base's above- and belowground facilities was said to extend into the billions of dollars. The conspiracies began in reference to who was behind the quakes and why they were supposedly executed so closely to the base, disrupting underground activities said to include everything from chemtrail storage, child sex-trafficking tunnels and holding areas, nuclear weapons storage, and alien technology. This underground facility has been dubbed Area 52 by many in the conspiracy field.

 This underground facility [China Lake Naval Base] has been dubbed Area 52 by many in the conspiracy field.

The area is prime for seismic activity, but anytime such a secretive place is affected, the questions begin. Was this a Russian attack of revenge or terrorism? Was our own government trying to hide something before a whistleblower had a chance to expose it? Was it part of a test on behalf of those involved with a greater agenda to cull the human population? The nefarious activities alleged to be occurring beneath the ground at China Lake Naval Facility even included a weather warfare operation, a mind-control unit, and a directed-weapons production facility. That last claim led to concerns that the military made an error while testing a DEW, creating a far larger quake than anticipated and destroying half its base in the process.

For scientists, there were plenty of clear signs that two large quakes had struck the northwest trending fault line in the area, including rupture signs, and that it was nothing more than nature doing what nature does.

After the quake activity and subsequent destruction, this author made note of the absence of chemtrail activity for approximately six months before it started up again. Coincidence? Or was China Lake part of a storage facility for chemtrail toxins, maybe even the touching-off point for planes to fly out and return to? Well, suffice it to say that if those lines in the sky were just regular contrails, and said author lives in an airport flight path, why did all contrails cease to exist for about six months?

Where there is smoke, there might be the beginning of a fire, an existing fire, or the end of a fire.

Speaking of fires, the recent horrific outbreak of wildfires in Australia has also seen its share of DEW conspiracy theories.

Australia Burns

Deadly outbreaks of wildfires occur all over the world. Conspiracy sites abound with posts and pictures of suspicious fires in Greece, Sweden, France, Norway, Asia, British Columbia, and Russia that began around 2018. But none is more devastating than the Australian wildfires of 2018–2019.

Beginning around the same time as the California wildfires, the entire continent of Australia was on fire as the flames destroyed homes and killed people and a shocking amount of wildlife before they were contained. The fires seemed to sweep across the continent at an unprecedented rate. Government officials and scientists agreed it was due to climate change and a hotter, dryer environment ripe for uncontrollable burning.

These were "bush fires," but the conspiracy-minded began whispering of a controlled effort and a sinister agenda behind what they claimed were DEW attacks. Again, interviews with firefighters and strange photos began appearing on social media, YouTube, and various websites, followed by the usual backlash from the mainstream media calling anyone who questioned the official explanation a conspiracy nutjob. The problem was, some media and Australian government spokespeople were not forthcoming with proper information the public felt they could trust simply because they didn't yet have all the facts. Word first arose of climate change as the cause. Then it was falsely reported that these were a concerted effort of a number of arsonists. It got to the point where one didn't know the difference between the official and conspiracy explanations.

Wildfires in Australia killed much of the wildlife, including kangaroos, koalas, and rare marsupials that are now threatened with extinction.

The same questions that arose during the California wildfires arose in Australia, with fingers pointing to "climate change denial" on the official side and fingers pointing to "strange burn patterns and inconsistencies with natural wildfires" on the conspiracy side. At the heart of the conspiracy side was the use of targeted DEWs to start the fires. What wasn't so clear was why. The Agenda 21 conspiracy was at the forefront with claims this was a test of DEWs to destroy or cull an entire populace. Others suggested there was some strange use of the fires to force—here we go again—a "high-speed rail program." Others still felt it was nothing more than an ill-equipped country facing a massive assault from Mother Nature herself and a harbinger of what was to come when the climate got hotter and drier.

The confusion continues today.

The bottom line may not be whether or not there is any truth to these conspiracy theories but that there is a global rising distrust of anything our media, government leaders, and politicians tell us. Whether you're in California or Australia, and regardless of your political stance, outright distrust of the powers that be is what fuels the drive to come up with alternative explanations for the information we are given via mass media. We know we've been lied to, deceived, harmed, manipulated, and worse by our governments, corporations, military, and religious leaders all over the world, and certainly in the United States, so it shouldn't be too hard to figure out why these conspiracy theories continue to proliferate.

Cell Phone Dangers

Remember a time when cell phones didn't exist? Most people don't, or they refer to it as the "distant past when dinosaurs roamed Earth." Cell phones are now owned by more Americans than cars or houses. Statistics show that in 2017, 67.3 percent of U.S. citizens had a smartphone. Today, that is estimated at over 72 percent and rising. Cell phone sales were about thirty-three billion dollars in 2012. Today, they are close to seventy-eight billion dollars per year, despite reaching saturation points. No doubt, many of these new sales come as younger children are given cell phones to stay connected to Mommy and Daddy or as distractions to keep them busy!

Cell phone usage is also at an all-time high as more people use social media and access the internet via their phones rather than older methods of desktop computers, laptops, tablets, and those ancient things called telephones and pay phones. We are attached to our cell phones, afraid to leave home without them, unable to find our way around town without them, and crippled with anxiety when we don't have automatic access to the world right in our own pockets or pocketbooks.

We are addicted to cell phones, but few of us really think about the potential harm they may cause to our brains and bodies

Before cell phones, people used to socialize with each other rather than stare at glowing screens. More than just annoying and possibly rude, cell phones might pose a danger to our health as well.

from overusage. That is something the cell phone industry does not want anyone thinking about. Cell phone manufacturers, wireless carriers, and others in the vast and growing industry want to make sure that any "medical" research on cell phone usage is skewed in their favor, proving that continuing and up-close use of the radiation these phones give off is not harmful in the slightest.

What happens, though, when you look at the research that isn't paid for or sponsored by the industry? Cell phones have only been around for a couple of decades or so, which means there are not a lot of long-term studies available as to how our brains and bodies react to the constant microwave radiation of holding a cell phone up to our heads. Yet a rising number of studies are finding direct links between increases in particular types of brain tumors and cancers and increased or long-duration cell phone use. When looking at this research, it always helps to follow the source to the money and the money to the source.

One telling note: all the way back as far as 2008, the CTIA–The Wireless Association was asked to attend U.S. congressional hearings discussing the possible dangers of cell phone usage. It refused, claiming the "science was settled." Remember that phrase because we will be hearing that again later when we talk about deadly vaccines and pharmaceuticals. Yet not only does every cell phone package insert contain instructions to not hold the phone within two-tenths to six-tenths of an inch near the body to limit radiofrequency (RF) exposure to under the federal limit (why is there a limit to begin with if completely safe?), but one of the first major studies by the World Health Organization's (WHO) International Agency for Research on Cancer in 2009, published later in the *Journal of Clinical Oncology*, found that cell phone radiation was "possibly carcinogenic to humans." That's a far cry from "the science is settled," and unfortunately for the cell phone industry, the news gets worse from there.

  Dr. Devra Davis, founder and president of the Environmental Health Trust … documented studies at the time that all led to one conclusion: most users were totally unaware of the dangers, which Davis considered a mistake.

In her 2010 book *Disconnect: The Truth about Cell Phone Radiation, What the Industry Has Done to Hide It, and How to Protect Your Family*, Dr. Devra Davis, founder and president of the Environ-

mental Health Trust, wrote about the "immense disconnect between the common public and medical opinion about cell phone
safety and what the data actually said." Her book documented
studies at the time that all led to one conclusion: most users were
totally unaware of the dangers, which Davis considered a mistake.
That divide today seems even larger as people have become so dependent on their cell phones, they go into a deep state of denial
when presented with information suggesting they may be compro-
mising their health.

Davis found that there are many myths and mistaken beliefs
about cell phone use, too, that should be cleared up:

▲ Sleeping with a cell phone near your head when
turned off is *not* OK. Even when off, cell phones emit
electromagnetic radiation. Put the phone on "airplane
mode" if you must keep it on to shut down the trans-
ceiver, and keep it several feet away from your bed,
especially your head area.

▲ Children are *not* safe when using cell phones. Their
brains are still growing, and their skulls are thinner,
meaning they are prone to absorb more radiation than
adult brains. Keep their use minimal, and do not let
them hold the phone to their ears.

▲ A weak signal does *not* mean you're safer. In fact, the
weaker the signal, the stronger the signal your cell
phone puts out to connect. Therefore, you're exposed
to more radiation. Wait for a stronger signal or use a
landline.

▲ Using one side to talk on a cell phone is *not* better.
Switch sides to spread out the exposure and limit the
harm, but no matter which side you use, hold the
phone a few inches from your ear.

▲ Keeping your phone in your pocket is *not* safe because
it means your phone is near your body for prolonged
periods of time. Keep it in a purse or case away from
the body if possible.

Sounds simple, yet look around and see how many cell phone
users don't follow these rules. Including yourself? We have become
so spoiled and so addicted to cell phone usage, we ignore any dangers,
or we make changes for a short period of time, only to default back

to dangerous usage. We might even engage in a little information bias by only choosing to believe the research studies in major publications or news sites that say cell phones pose no threat of harm, never realizing that these publications and news sites make millions of dollars off advertising from cell phone companies, which means the "research studies" will always be skewed in favor of the industry. Follow the money to the source and the source to the money.

Davis has continued to research cell phone harm and the risks especially to pregnant women, their unborn fetuses, and small children. She has warned for years about altered DNA, altered brain metabolism, learning disabilities, and compromised spinal cords from cell phone radiation exposure in prenatal animal studies. Children are at risk because their brains contain more liquid, which we know absorbs radiation (think of how a microwave oven cooks water molecules in food) in far greater amounts than the average adult, and the risk of brain cancer is about four to five times higher for teenage children who began using cell phones when much younger.

Cell Towers

It isn't just the phones themselves; the cell phone towers are linked to adverse effects as well. In a 2013 article appearing in the *British Medical Journal*, it was found that being within 1,600 feet of a cell phone tower showed increased reports of vertigo, lack of concentration, irritability, insomnia, and loss of appetite. If we consider the fact that there are over 200,000 cell phone towers spread throughout the United States, and that is a lowball figure, it's not hard to see how our entire sky is blanketed with the invisible and harmful radiofrequency waves. Wait until we get to the new rollout of 5G.

Since then, numerous studies have documented the negative effects of public exposure to RF waves. Even though these waves are not as powerful as gamma rays or ultraviolet rays, we are just learning the dangers of low-level, ongoing exposure despite the constant bombardment from the cell phone carriers that there is nothing to be concerned about. The public is now exposed to over 100,000,000

Cell phone towers surround us these days, emitting radio frequencies that, according to studies, have negative effects, including neurological harm.

times more electromagnetic radiation than previous generations, according to Jim Marrs in *Population Control*. He writes, "If you can make a cell phone call, you're in an area saturated with microwave radiation."

The intensity of that radiation decreases with distance and is all but gone at the twenty-one-mile point, but most of us can look outside and spot a cell phone tower well within a twenty-one-mile radius. The electronic pollution put out by these towers is doing neurological harm and possibly causing cancer, especially to those who live close to a cell tower base station. Because the cell tower signal strength is determined by the inverse square of the distance, Marrs reminds us, a person living twice as close to a cell tower is bombarded with four times the radiation.

Cell phone and microwave radiation has the most adverse effects on the brain, with increases in the aggressive and deadly glioblastoma. One damning study was conducted in France yet mirrored findings in many American studies. The French study involved the Santé publique France, along with the FRANCIM cancer registries, the Hospices Civils de Lyon, and the Institut National du Cancer. These organizations published national cancer incidence and mortality estimates in metropolitan areas of France from 1990 to 2018. The studies were based on modeling of observed incidences up until 2015 and projections up to 2018.

 In 1990, there were 823 cases of glioblastomas.
In 2018, there were about 3,500.

The first volume of the study looked at solid tumors and found that between 1990 and 2018, the overall rate remained stable in men but increased in women. However, the annual number of new cases of the deadly glioblastoma increased more than fourfold in both genders. Over the course of 30 years, the number of cases multiplied by four. In 1990, there were 823 cases of glioblastomas. In 2018, there were about 3,500. According to the Santé Publique France (Public Health France), this mirrored findings in the United States as well as an Australian study done from 2000 to 2008. The conclusion was that the factors that played a role in the increased rates of glioblastomas were brain radiation therapy, pesticide exposure, and increased exposure to electromagnetic fields.

Dr. Marc Azari, president of the Phonegate Alert Association, told Global Research Online, "Over the last 2 decades, nearly

50,000 people have been affected in France by this extremely aggressive brain tumor, which has a very high mortality rate. It was also during this period that mobile telephony exploded, and industrialists knowingly overexposed us to the waves of our mobile phones. This industrial and health scandal has a name, the 'Phonegate'!"

Phonegate is the perfect name because the WHO and the Centers for Disease Control and Prevention (CDC) have both been accused of showing cronyism and favoritism of studies financed by the cell phone industry and refusing to firmly recognize the dangerous links between cell phones and EM radiation and hazards to human health, including cancer.

CDC Cover-up

In 2014, the CDC finally stepped up and posted this notice on its website: "We recommend caution in cell phone use." But within a few weeks, it had taken the warning down, along with a section about the potential risks of cell phone use to children. In 2016, it posted: "Some organizations recommend caution in cell phone use. More research is needed before we know if using cell phones causes health effects." The *New York Times* got hold of over 500 pages of internal documents suggesting disagreements about the use of cell phones between CDC scientists and key agency officials.

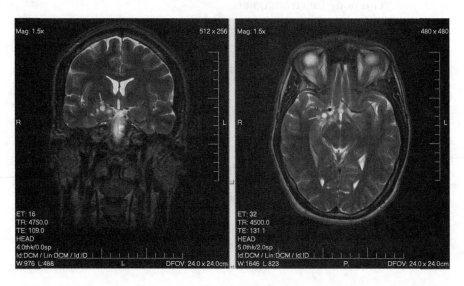

MRI scans showing a glioma inside a human brain. Gliomas are tumors that affect the brain or spine and are frequently malignant. There appears to be a connection between cell phones and an increased chance of gliomas.

The CDC reportedly backpedaled and, after years of debate and the use of focus groups to test out new language, still did not go so far as to say caution should be used. Why? There was plenty of research showing links between microwave radiation and brain tumors, including studies done by Dr. Lennart Hardell, professor of oncology at the University of Örebro in Sweden, along with statistician Michael Carlberg, showing case-controlled studies of Swedish patients diagnosed with malignant brain tumors from 1997 to 2003 and from 2007 to 2009.

While most scientists have asserted that normal cell phone use has not been shown to cause cancer (because the radio frequencies are nonionizing and do not damage cells), a 2018 article published in *Scientific American* reported a disturbing study that suggests otherwise. The National Toxicology Program at the National Institutes of Health ran an experiment involving 3,000 rats exposed to cell phones that caused some of them to develop schwannoma, a cancer of the Schwann nerve cells. Experts from top universities took part in the 2013 "Cell Phones and Wi-Fi—Are Children, Fetuses, and Fertility at Risk?" conference and concurred that there is a direct relationship between the duration of cell phone use and sperm count decline. They said that sperm counts are cut in half in men who carry cell phones in their pants pockets for more than four hours a day.

The Body Electromagnetic

In the November 30, 2015, issue of *Waking Times*, Barrie Trower, a physicist and former military expert on microwave radiation, said in "Electromagnetic Radiation and Other Weapons of Mass Mutation" that "there are no safe levels of radiation." Trower reports that the brains of those surrounded by microwaves, RF frequencies, and the like in control rooms, military, IT, financial, or other environments are not going to function normally because of the constant bombardment of electromagnetic radiation (EMR). But this also includes smart homes with appliances, televisions, computers, baby monitors, cameras, lighting, heating, and other sources. Brains just don't function the same when exposed to these kinds of stimuli.

 Electromagnetic hypersensitivity syndrome is on the rise, according to several studies.

Could this explain the overall malaise people complain about—such as their inability to focus, concentrate, or achieve a particular goal? The exhaustion and depression that plague people without explanation? Lack of sleep? The evident hostility of people on social media and on the roadways?

Electromagnetic hypersensitivity syndrome is on the rise, according to several studies. A 2008 Australian study showed the prevalence of this syndrome rose from 1.5 percent in 1994 to over 3.5 percent in 2008. A 2006 German study showed a syndrome rate of 9 percent, and Taiwan reported a rate of 13 percent in 2011! The more we become surrounded by EMR sources, the more this syndrome rises. Some of the symptoms reported are:

▲ Rashes on the skin, burning or itching sensations, or tingling, especially around the head and chest area

▲ Headache and earache

▲ Vertigo and dizziness

▲ Insomnia

▲ Panic attacks

▲ Chest pain and heart issues

▲ Buzzing or vibration sensation throughout the body

▲ Seizures

▲ Fatigue and muscle aches

▲ Paralysis

The WHO reported back in 2005 of the health problems associated with EMF exposure, stating some people were literally debilitated by it. Now, with cell phones added to all the other existing sources, it's no wonder people are always complaining of the above symptoms or just generally feeling sick and tired. This exposure also contributes to Alzheimer's, autism, depression, miscarriages, anxiety, and cardiac issues such as heart tumors from RF radiation from 2G and 3G cell phones. We are now seeing the coming of 5G!

In 2011, the International Agency for Research on Cancer, which is part of the WHO, declared cell phones a Group 2B—a possible carcinogen. New research shows that they cause not only cancer but also damage to the mitochondria of cells. Electromagnetic fields (EMFs) trigger peroxynitrites, which damage mitochondria and contribute to a variety of chronic diseases. While the body has the capacity to repair damage, thanks to the presence of a family of enzymes called poly (ADP-ribose) polymerases, the enzymes cannot function properly because of the EMF exposure and therefore cannot repair the DNA. This leads to cell death.

More Negative Studies

The heat emitted as radiation is not the only problem we face from overexposure to cell phones and other devices. Dr. Martin Pall, professor emeritus of biochemistry and basic medical sciences at Washington State University, found more than 20,000 published scientific studies that showed "significant biological effects at exposures well within safety standards, including about 4,000 that discuss thermal effects," according to an article on Mercola.com titled "Guys: Your Cellphone is Hurting Your Sperm." In addition to the heating of tissues in the body, cell phones are also causing risk from the frequencies used, the erratic pulsing and modulation of the signals, the magnetic fields of the batteries, and radiation penetration into the head. U.S. safety standards only take into consideration the harm caused by radiation heating, thus leaving out a whole host of damaging effects the public is unaware of.

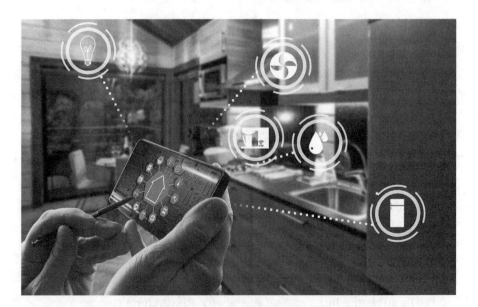

Many of the electronics inside your home are emitting electromagnetic radiation that can damage everything from reproductive to neurological systems in your body. They could also lead to cancerous tumors.

Dr. Martin Beck, an expert on cellular and molecular effects of electromagnetic fields, gave a speech called "The Health Effects of Electromagnetic Fields" at the Commonwealth Club of California. The speech, cosponsored by ElectromagneticHealth.org, was a basic explanation of how the structure of human DNA, the coil of coils as Beck put it, is especially vulnerable to electromagnetic fields. He expressed concern for how the long-term exposures to such fields do significant harm to our genetic material, and he pointed to a body of scientific research stating these harmful effects were real.

DNA mutations and damage have been reported not just for children and men's sperm counts but also for fetuses in the womb. Fertility rates in women are also adversely affected by long-duration cell phone use, in addition to other types of exposure to electromagnetic energy, whether from smart appliances in the home or what is known as dirty electricity, including CFL light bulbs, cell phone antennas, power supplies for portable computers, phone chargers, light and fan dimmer switches, and other electronic devices that use transformers to convert voltage. Dirty electricity is the result of high-frequency transients on electrical wiring specifically. Dirty EM fields are invisible but are associated with a host of illnesses, according to *Dirty Electricity: Electrification and the Diseases of Civilization* by epidemiologist Samuel Milham, MD.

Milham writes that dirty EM fields can cause diabetes, cancer, heart disease, and autoimmune diseases, which are much less prevalent in populations where there is no electricity or less of it. Even suicide is on the list of the negative effects. So, it isn't enough to just blame cell phones. All sources of electromagnetic radiation are the problem. To minimize the risks, the following is suggested:

▲ Reduce or eliminate the use of wireless devices. If using a phone at home, use the older types operating at 900 MHz, which still give off EM fields but don't broadcast those fields when not in use.

▲ Reduce or eliminate cell phone use in general, and turn it off when possible.

▲ Never let a child use a cell phone for entertainment. Only teach them how to call 911 in an emergency.

▲ Do not use cell phones in areas of low reception and weak signals.

▲ Do not carry wireless devices or cell phones near the body or keep them near the body on a desk or table.

▲ When using headsets, choose the wired style, but make sure they are well shielded.

▲ Keep your laptop off your lap!

▲ Replace Wi-Fi routers with Ethernet cables.

▲ Stand at least three feet away from a microwave oven when in use.

▲ Avoid electric heating pads, blankets, and other bedding items.

▲ Change all your fluorescent light bulbs to incandescent light bulbs.

▲ Ask your utility company to give you a wired smart meter, not a wireless one.

▲ Use newer flat-screen televisions or computer monitors, as they emit less EMR than the older models.

Doing just a few of these things can go a long way in cutting your risk of exposure to levels of EM radiation and reduce the possibility of cancer and other diseases. Most households have ditched their landlines for cell phones only, but it might behoove you to use a landline when calling from home and save the cell phone for when you're away. At the very least, it's a good idea to keep the cell phone as far from your head and body as possible, to not allow your children to use them as a toy or distraction, and to do a little research into the potential dangers rather than blindly trusting the advertising and "scientific studies" sponsored by the very companies that benefit financially from you staying ignorant about the dangers.

 People who used their cell phone this often (about seventeen minutes per day for ten years) had a 60 percent increased risk of developing brain tumors.

On November 2, 2020, the *International Journal of Environmental Research and Public Health* published an open-access article titled "Cellular Phone Use and Risk of Tumors: Systematic Review and Meta-Analysis," which analyzes data from forty-six case-control studies that looked at the relationship between cell phones and cancer. The new meta-analysis revealed "significant evidence linking cellular phone use to increased tumor risk, especially among cell phone users with cumulative cell phone use of 1,000 or more hours in their lifetime." People who used their cell phone this often (about seventeen minutes per day for ten years) had a 60 percent increased risk of developing brain tumors.

The Mayo Clinic and the National Cancer Institute also cited studies linking heavy cell phone use and increased incidences of gliomas and salivary gland cancers. The American Cancer Society stated radiofrequency radiation is rated *"possibly carcinogenic to humans"* by the International Agency for Research on Cancer (IARC). Wireless headphones also may be a problem, but they emit lower radiofrequency energy than cell phones. Our young people who wear their headphones all day and night, and are on their phones hours at a time, must know the risks.

Microwave Oven Dangers

Just about everyone has a microwave oven, but few people understand the way these appliances work. Microwaves use a specific set of electromagnetic wave frequencies to agitate the water molecules found in food. As the water molecules become increasingly agitated, they vibrate at the atomic level and generate heat to cook the food. They literally cook the food quickly, which is why they are so desirable in our fast-paced world.

On the FDA's own website, fda .gov, there is a warning about using microwaves, urging users to take precautions handling hot food and beverages that could scald when removed right from the microwave, but there is also a vague warning about how microwave radiation can heat the body tissue the same way it heats food and that exposure to high levels of microwaves can cause painful burns, especially to the eyes and testes, which have relatively little blood

If microwaves can cook food and boil water, what might they do if you have an unsafe oven that leaks radiation? Even when a microwave oven is in good repair, you should not stand close to it while it is in use.

flow to carry away excess heat. However, the FDA makes sure to add that these kinds of problems only occur after exposure to large amounts of microwave radiation. Yet we must keep in mind the FDA is beholden to the food industries, many of which thrive on microwaveable products.

Microwave ovens produce nonionizing radiation, which doesn't cause the genetic damage of ionizing radiation from nuclear plants and the like. But that doesn't mean there are no dangers associated with nonionizing radiation. Although microwaves are supposed to be built to contain the radiation, there is often leakage, enough to prompt the FDA to set strict limits on microwave manufacturers, especially when it comes to doors preventing radiation leaks. It's a small amount of leakage if any, according to manufacturers, and perfectly safe unless the microwave oven is old or somehow defective or the seal on the door is compromised. The manufacturers do warn against standing in front of a microwave oven in use, and many health-oriented websites suggest standing a few feet away at the very least, in case of EM-energy leakage.

So, who do you believe? The actual microwave oven industry and its lobbyists or the alternative health sources that report on studies showing that microwaves are unsafe and cause cancer? The bottom line is you must decide whether you can trust food that is cooked in an unnatural manner for the sake of speed or determine whether the machines meant to cook it are safe. That's an individual call you must make when buying any appliance that deals with or emits EMR.

But it's not just the EM energy given off during microwave use, it's the friction caused during the heating of the food that can cause damage to surrounding molecules by ripping them apart and causing deformities in the food itself. In 1991, Swiss scientists Hans Ulrich Hertel and Dr. Bernard H. Blanc of the Swiss Federal Institute of Technology and University of Biochemistry tested microwaved food. Test subjects were given preparations of milk and vegetables in their raw state, heated conventionally, or microwaved. Then blood samples were taken before consumption, during, and after.

Those who ate the nuked food showed a decrease in hemoglobin, an increase in cholesterol values, and high white blood cell counts, all signs of an inflammatory response. Higher white blood cell counts results in leukocytosis, a sign of pathogenic effects such as poisoning and cellular damage. Other studies have shown an increase in known carcinogens called dioxins from microwaving food. The dishes used in the microwave can also leakbiphenol A (BPA), which is linked to breast cancer. One study done in 2010 at Trent University concluded that there is ample evidence that microwave frequency radiation harms the heart. As far back as 1989, Dr. Lita Lee reported in the medical journal *Lancet*, "Microwaving baby formulas converted certain trans-amino acids into their synthetic cis-isomers. Synthetic isomers, whether cis-amino acids or trans-fatty acids, are not biologically active. Further, one of the amino acids, L-proline, was converted to its d-isomer, which is known to be neurotoxic (poison) to the nervous system and nephrotoxic (poisonous to the kidneys)." In 1991, an Oklahoma lawsuit involved blood heated in a microwave in a hospital that killed a woman after her transfusion. It was later determined that microwaving alters the structure of blood.

 A study on human breast milk conducted in 1992 and published in the April issue of *Pediatrics* showed microwaved breast milk promoted dangerous bacteria, reduced key antibodies, and cut down lysozyme activity.

A study on human breast milk conducted in 1992 and published in the April issue of *Pediatrics* showed microwaved breast milk promoted dangerous bacteria, reduced key antibodies, and cut down lysozyme activity. Not only that, but the heated milk also lost 96 percent of its beneficial immunoglobulin-A antibodies. The study concluded that microwaving food reduces and reverses the nutritional benefits above and beyond the harmful effects of the heating process.

Microwave ovens make life easier, but it's much better to cook food conventionally to avoid adding even more EMR to your environment, not to mention the damage it causes to food itself once you've eaten it. Let's not even discuss the chemical-laden processed foods people often buy as microwave meals. That's a whole other section. You can obviously cut down the potential damage and harm by using the microwave a lot less, standing as far away from it as you can, and not allowing children to use the ovens. Of course, you must make sure the dishes, glasses, and plates you put in the microwave say they are microwave safe and never put foil or plastic in the ovens.

The best way to avoid the exposure to EMRs from microwaves is to just not use them, but that is easier said than done in this busy/crazy world.

Buyer beware.

A hair dryer can emit about 400 mG of electromagnetic radiation during one usage, which is about a quarter of the maximum (2,000 mG) considered to be a safe dose.

EMF Dangers

An area of great concern occurs in our bedrooms, where we spend a large chunk of time sleeping. EM fields from electric blankets, waterbeds, electric clocks, and computers all contribute to more bombardment of EMR, even in small amounts. Keeping electric heaters and heated blankets on low settings can help, as can moving clocks and computers as far from the bed as possible, at least a yard away. If the blanket emits less than 2,000 mG (mG stands for miligauss, a unit of measurement for magnetic induction), it should be somewhat safe, although the fact that people keep them on all night and wrapped around their bodies opens them up to possible biological stress and overexposure.

Everyday items such as electric razors and blow dryers also contribute to the amount of EMF in the home, as do all appliances, so obviously, the less amount of time spent using them in close proximity, the better. When turned on, an electric razor or hair dryer can emit up to 400 mG, and most experts suggest that is way too high. Think about this: the electric clock on your bedstand emits about 5 to 10 mG, which is the equivalent of an overhead electric power line!

The kitchen is the most dangerous place of all, with numerous appliances running on electricity, and many of the new smart appliances emit stronger EMF than the old-fashioned types. When spending time in the kitchen, it's best to keep things unplugged when not in use. Another good suggestion is to have the family eating area separate from the cooking area with appliances.

Smart meters operate wirelessly and give off strong and short pulsed microwave radiation, which many citizen action groups claim is dangerous to human health. Smart meter signals can often exceed the ranges the gas and electric companies claim are safe, and sometimes, that upper range can hit 190,000 daily pulse bursts, according to citizen watchdog groups. The effects of this pulsed radiation on human biological systems have not been studied, and citizen groups have been successful in showing correlations with ill health of people using them, enough to prompt some utility companies to adapt smart meters to transmit on a schedule of once per day instead of once per minute, although there are still exposure concerns from smart meters owned by neighboring homes.

If you are living in an older neighborhood or a more rural area, you may have exterior power lines to deal with. These high-voltage transmission lines use high-voltage direct current (HVDC) to transmit power from the local generating station and emit powerful electromagnetic fields that have been linked to cancer and other diseases in birds, animals, and humans. Homes near or below power lines are much less desirable to buyers, as well as those near transmission substations and transformers. Transformers can emit an EM field up to one-quarter of a mile away.

There are lower-voltage power lines in many neighborhoods, and their risk varies with voltage strength and whether or not they are above or below ground. Buried power lines are thought to be safe, but that's not entirely true. While burying power lines below ground eliminates the electrical component of the EM field, the magnetic component remains, which easily passes through the ground into homes and living things.

Studies showing the negative biological effects of power line exposure go back over twenty years, with increases in brain cancers, birth defects, lymphoma, ALS, and miscarriages being recorded. Children living in homes within 650 feet of power lines and sources are at a great risk of developing leukemia, upward of 70 percent more than children living 2,000 feet away or more, according to a 2005 study published in the *British Medical Journal*. People living within 328 yards of a power line up to the age of fifteen are three times more likely to develop cancer as adults, and according to Dr. David Carpenter, professor of environmental health sciences at SUNY at Albany, an estimated 30 percent of all childhood cancers can be linked to power line exposure.

 Children living in homes within 650 feet of power lines and sources are at a great risk of developing leukemia, upward of 70 percent more than children living 2,000 feet away or more....

According to the National Cancer Institute, the nonionizing radiation from power lines does not cause cancer. The WHO and other national agencies reviewed studies done during the 1980s and concluded there was no risk from exposure to EMFs and that high exposure was probably associated with wrong wiring configuration inside the home and not outside. But that doesn't explain how much more powerful energy fields emitted by power lines above or near the home would *not* affect someone when the implication is they do affect someone inside the home.

So, there are both pro and con studies regarding external power lines and EMF exposure. Next time you're buying a home, do some research if there are power lines present, and err on the side of caution.

5G: FASTER. CHEAPER ... DEADLIER?

In our quest for cheaper and faster cell phone service, smartphone technology goes through "generations" that indicate the type of technology used. Cell phones started off with 1G analog technology in the 1980s, then evolved to 2G digital, then to 3G, which provided higher-speed networks. Then came 4G and 4G LTE (Long Term Evolution), which improves upon the speed of 3G for smartphones, tablets, and wireless hot spots.

Today, we have the rollout of 5G, which includes both LTE and VoLTE (Voice over Long-Term Evolution) and is the cutting edge of cell phone service and speed. 5G is said to be over 100 times faster than 4G service, and with the abundance of 5G towers erected in cities and towns the country over, as well as more in the works, the entire sky will be blanketed with microwave radiation, all for the sake of a faster phone call or text.

While those in the 5G industries touted all of the advancements this new generation of technology would bring, basically high-efficiency data transfer, skeptics claimed the customers would barely notice the difference in speed. But they might notice the physiological effects of the 5G onslaught of thousands of towers sending nonstop, low-level radiation into just about every nook and cranny from major cities to rural areas lucky enough to get towers constructed in their areas.

A 5G cellular network antenna looms over a city full of unsuspecting citizens as to the potential dangers of the EM waves it emits.

In March 2018, a scientific peer review of the U.S. National Toxicology Program study into mobile phone radiation and health found clear evidence that radiation from cell phones caused cancers of the heart tissue, the brain, and the adrenal glands. But the U.S. media was silent on the issue and mobile phone manufacturers and carrier services were, too, instead launching a PR campaign to counteract the negative studies and mislead journalists, consumers, and government officials by claiming cell phones had no links to cancer.

In December 2019, a large and broad coalition of scientists, researchers, doctors, and advocates signed a National 5G Resolution letter and sent it to President Donald Trump demanding a moratorium on 5G until further studies on the potential hazards to humans and the environment could be fully undertaken by independent scientists who were not beholden to the telecom industries. The letter referenced published studies that clearly demonstrated harm to human health, bees, the environment, wildlife, trees, and birds from existing technology and showed that the increase needed to accommodate 5G would be even more disastrous.

It also emphasized the dangers to children, as per this statement by participant and pediatrician Toril H. Jelter, MD, who was a presenter at the conference on EMF case studies. She specifically studied children and commented on their improved health when their exposure to wireless technology was removed or reduced: "The children are our future. The scientific evidence has been clear for decades and now America has an opportunity to lead the way.... It is my impression that health effects of wireless radiation go misdiagnosed and underdiagnosed for years. Parents, teachers, and physicians need to know that hardwiring internet, phone, and TV is a healthier option for our children."

 A petition signed by 26,000 scientists and countless others who opposed 5G was also sent in 2018 to the United Nations, WHO, the European Union, the Council of Europe, and the governments of all nations.

A petition signed by 26,000 scientists and countless others who opposed 5G was also sent in 2018 to the United Nations, the World Health Organization (WHO), the European Union, the Council of Europe, and the governments of all nations. The first paragraph stated:

> We the undersigned scientists, doctors, environmental organizations and citizens from (...) countries, urgently call for a halt to the deployment of the 5G (fifth generation) wireless network, including 5G from space satellites. 5G will massively increase exposure to radio frequency (RF) radiation on top of the 2G, 3G and 4G networks for telecommunications already in place. RF radiation has been proven harmful for humans and the environment. The deployment of 5G constitutes an experiment on humanity and the environment that is defined as a crime under international law.

The executive summary that followed laid out the case for the dangers of 5G and how its involuntary nature violates international ethics and morals laws and codes. Yet as of the writing of this book, 5G was rolling out in more countries despite these warnings and urgings for more scientific research.

The Federal Communications Commission (FCC) had stated that over 800,000 antenna sites were required to deploy 5G in the United States alone, with over 5,000,000 by 2012 globally, and that wireless antennas were already being rapidly constructed using streetlights and utility poles (sometimes being disguised as fake trees) in neighborhoods all over the country. This would dramatically increase the amount of radiofrequency and electromagnetic field exposure on top of the pre-existing 2G through 4G infrastructure, and this didn't include the hundreds of satellites placed into orbit by hydrocarbon rocket engines, which contribute to atmospheric pollution, all without any premarket safety tests. This completely ignored the 2015 studies on behalf of over 230 scientists into the health effects of non-ionizing EMFs done in 41 different nations as part of an international appeal to the United Nations calling for protections from EMF exposures at low levels.

One of the main concerns involved the reliance of 5G on the bandwidth of the millimeter wave (MMW), known to be able to penetrate one to two millimeters into human skin tissue, meaning that sweat ducts in human skin act as an antenna when in contact with MMWs. The symptoms of exposure include burning sensations, pain, eye problems, altered heart rate, arrhythmias, and suppressed immune functions. This new type of environmental pollution has been banned in some cities and countries, and more are signing on to keep it out of their neighborhoods, but the telecom industries are powerful and have at their disposal politicians, lobbyists, and corporations with major clout in Washington.

In the May 29, 2018, edition of the UK *Daily Mail*, University of California at Berkeley public health professor Dr. Joel Moskowitz said, "The deployment of 5G constitutes a massive experiment on the health of all species." He also said that, since 2009, hundreds of journalists had interviewed him about the effects of cell phone health, but very few had asked about 5G's health effects. Moskowitz believes the rush to get 5G up and running shows how the telecom companies have convinced U.S. policymakers and the public that it is a global race for superior technology with China and other countries and unless we are first to deploy, we won't get the benefits.

Starlink Initial Phase

1,584 satellites into 72 orbital planes of 22 satellites each

The SpaceX corporation is deploying Starlink, a network of satellites in low Earth orbit for use in communications. This graphic shows Phase 1 in which nearly 1,600 satellites are placed in 72 orbits. Starlink should be completed by 2027.

The telecom industry insists 5G is a necessity to make the internet work better, but they don't mention the billions of dollars in profits they hope to make from the newest generation of technology. Nor do they mention the use of 5G technology and smart meters as a part of a growing surveillance and data harvesting plan that is invasive and intrusive but is also big business. Nor do they mention that 5G emits the same frequencies into the air that riot control weapons use.

In "5G Danger: 13 Reasons 5G Wireless Technology Will Be a Catastrophe for Humanity," published in *Global Research*, Makia Freeman writes that the 5G danger cannot be overstated. The dangers include:

▲ Damage to sweat ducts from the broadcast frequencies that could alter the human skin itself.

▲ Dangers from premature aging and injury to the body, the brain, fertility, hearts, and DNA from the way EMFs activate the body's voltage-gated calcium channels (VGCCs).

▲ Damage from pulsed-wave radiation, which is far worse than continuous radiation.

▲ Damage from deep EMF penetration when cell phones are held close to the brain.

▲ 5G is a weapons system disguised as a consumer product, similar to long-range radar, phased-array radar, and directed-energy weapons.

▲ Close proximity to the 5G towers causes negative effects in humans, which is reportedly linked to hundreds of birds in the Netherlands falling to their deaths from the sky during a 5G test.

▲ The same frequencies used in 5G are used in riot control and crowd-dispersal directed-energy weapons.

▲ The MMW frequencies are both mutagenic (causing DNA damage) and carcinogenic (causing cancers).

▲ Insects, birds, and children are the most vulnerable to 5G frequencies because of their body size.

▲ 5G is planned to be deployed in space as a grid no one will be able to escape from.

▲ 5G is part of a larger agenda to control and surveil humans.

Many conspiracy theorists posit that the intention of 5G is to make everyone sick, tired, and ineffective to fight back against the growing Agenda 21 and Agenda 2030 plans to cull the human population and save precious resources for the elite few. Unlike chemtrails poisoning the skies, 5G can be excused as a way of increasing our ability to communicate and access information via cell phones and the internet and play upon the addiction of humanity to their "gadgets" even as they are being poisoned from above as a result.

Perhaps this war for our air and health is summed up by this. While more countries and cities come on board to ban 5G until further health studies are available, at a 2016 National Press Club speech, then–FCC chairman Tom Wheeler, who was once the head of a large wireless industry lobbying group, summed up the industry side by saying, "Stay out of the way of technological development. Unlike some countries, we do not believe we should spend the next couple of years studying.... Turning innovators loose is far preferable to letting committees and regulators define the future. We won't wait for the standards...." A callous disregard for human safety is the hallmark of the telecom industry's responses to calls for more studies.

> ⚠ A callous disregard for human safety is the hallmark of the telecom industry's responses to calls for more studies.

A study on the effects of 5G looked at two populations in the United States around the turn of the twentieth century: rural and urban. Up until around 1950, rural areas were not yet electrified but urban areas were. Epidemiologist Samuel Milham, the author of *Dirty Electricity: Electrification and the Diseases of Civilization*, analyzed mortality statistics between these two groups and showed a wide difference in mortality from heart disease, cancer, and diabetes. Interestingly, as the rural areas became electrified, the two curves merged.

In 2017, Mercola interviewed biochemist Martin Pall, Ph.D., who indicated excess oxidative stress resulting in mitochondrial dysfunction. As pointed out earlier, this would lead to cell death. So, cancers alone don't make up the main health risks of EMF, as the voltage in our bodies plays a significant role. Yes, we produce our own electricity, which allows our cells to communicate and do the basic things they need to do to assure our survival. EMF interference in this process introduces frequencies that alter and damage the very processes necessary for a healthy nervous system, brain, heart, and reproductive organs.

This damage will be highly amplified if 5G frequencies are all over the airwaves and we cannot escape them.

Another danger of 5G is that it is a global network, which would make it difficult to predict the weather and could also prevent satellites from detecting major storms that might affect airports, the Navy, and civilians in flood zones. The widespread coverage of frequencies would interfere with those used to detect changes in water vapor and other aspects of predicting storms and weather patterns.

So far, the FCC has sided with powerful industry voices in insisting there are no dangers and that nothing should stand in the way of progress and the rollout of the massive 5G onslaught. Skewed research studies, media manipulation, and heavily funded lobbying influence make it all but impossible for citizen action groups to be heard and acknowledged. Perhaps when things die at unprecedented rates, the powers that be and their money machines will take notice and demand the studies that should be going on now. Perhaps not. Hopefully, more countries and cities will seek to ban this technology until they are satisfied it is safe, although sadly, they will still be affected by the microwave radiation from towns and cities nearby.

Meanwhile, Mercola suggests some additional ways to remediate exposure to an electromagnetic field (EMF) in your home, such as using a battery-powered alarm clock; getting rid of wireless baby monitors; using your cell phone speakerphone instead of holding it up to your ear on long calls; placing routers inside a shielded bag when not in use; shutting off Wi-Fi at night or getting rid of it for good; adding a shield to your smart meter, which can reduce the radiation by a huge amount; and replacing your old microwave oven with a steam convection oven, which is much safer and just as quick.

The Coronavirus/5G Conspiracy

In 2019, the novel coronavirus known as COVID-19 broke out in Wuhan, China; by early spring 2020, it was labeled a global pandemic. As both case and death counts rose, it wasn't long before conspiracy theories arose, too, including one that linked the rollout of 5G in the very same area with the explosion of cases of the virus. The 5G implementation occurred in Wuhan on November 1, 2019, around the same time the first virus case was being reported by the

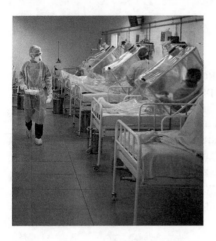

There has been considerable speculation as to the origins of the COVID-19 virus, but what if there is some connection with the 5G pilot program in Wuhan?

American media. Wuhan, the capital city of the province of Hubei, was one of the first pilot cities in China for 5G, with over 300 base stations set up since February 2018 and over a dozen experience centers completed in the cities of Xiaogan, Yichang, and Jingmen, where people can go and learn about 5G technology. By the end of 2019, Wuhan had over 10,000 antennas installed; as part of the 5G pilot project, by 2020 many were installed in Wuhan Vulcan Mountain Hospital as well as other area hospitals, which were ground zero for COVID-19 patients, to "assist in the battle against the virus," as claimed by Hubei Mobile and Hubei Unicom.

Coincidence? Strange timing? Or was this a coordinated attempt, as many believe, to get the hundreds of thousands of protestors off the streets of Hong Kong by "creating" a supervirus that spread like wildfire, then quarantining people and threatening those who knowingly spread the disease? And was the fact that a mandated vaccine was already being mentioned at the earliest stages of the viral outbreak part of the plan?

The *Diamond Princess* cruise ship with over 600 Japanese aboard was continuously streaming 5G at 60 Hz via the provider MedallionNet.

Meanwhile, there were reports that the virus was spreading in South Korea, considered the most heavily connected country on Earth, as well as in European nations where 5G was either already implemented or in the beginning stages, including England, Italy, and France. Italy was hit particularly hard early on by the COVID-19 virus. Interestingly, Italy was part of the 2016 EU 5G Action Plan and one of the first European countries to roll out 5G in late 2019 and early 2020, mainly in Turin, one of the country's first test spots as well as in Milan, Rome, Naples, and Bologna. Conspiracy theorists reported on the absolute lack of virus reports from third-world countries with no internet or cellular communications services and to the spread in the United States, where 5G towers are everywhere, as proof of their claims.

During the first few weeks of lockdown from COVID-19 in the United States, social media was filled with videos and pictures of telecom workers installing 5G towers near elementary schools and in cities that were all but abandoned at the time. Because there were no children at school or people in the streets, it seemed a rather stealth method of getting the technology up and ready without problematic inquiries and outright anger from concerned citizens and scientists.

Interestingly, in June 2020, a scientific study titled "5G Technology and Induction of Coronavirus in Skin Cells" was published in the *Journal of Biological Regulators and Homeostatic Agents*. The study was undertaken by several scientists working at the Department of Nuclear, Sub-Nuclear and Radiation Physics at G. Marconi University in Rome, Italy; Central Michigan University in Saginaw, Michigan; and the Department of Dermatology and Venereology, I. M. Sechenov First Moscow State Medical University in Moscow, Russia. The prestigious scientists involved showed that "5G millimeter waves could be absorbed by dermatologic cells acting like antennas, transferred to other cells and play the main role in producing coronaviruses in biological cells," according to the paper's abstract.

The study found that to produce viruses within a cell, such as coronaviruses, the wavelength of the external waves must be shorter than the size of the cell. Thus, 5G millimeter waves could be good candidates for applying in constructing viruslike structures such as COVID-19 with cells. The study is filled with discussions of DNA and the structure of cells as well as the influences of waves emitted from towers in 5G and other technologies the researchers stated "could have more effect on evolutions of DNA within cells." Their conclusion

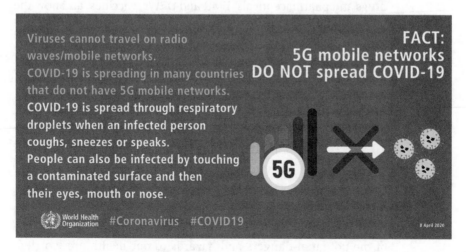

The World Health Organization released this graphic explaining that COVID-19 cannot be spread by 5G networks, but at least one scientific paper disagrees.

also stated that "a DNA is formed from charged particles and, by its motions, electromagnetic waves emerge. These waves produce hexagonal and pentagonal holes in liquids within nucleus and the cell. To fill these holes, bases are produced. These bases join to each other and can construct viruses like Coronaviruses."

The paper literally stated 5G and COVID-19 could be linked, with the influence of 5G waves not only creating the virus within the cells but possibly exacerbating symptoms as the symptoms of both EM exposure like 5G and COVID-19 were so similar. Frightening, indeed, and one of many reasons why hundreds of scientists, doctors, and researchers have been speaking out on the dangers of 5G ever since.

The truly terrifying thing is that the *Journal of Biological Regulators and Homeostatic Agents* retracted the paper one month later after the media blasted it as being false, flawed, and a conspiracy.

Could the effects of 5G cause people who were getting the virus, which has similar symptoms to the flu, to become sicker and more debilitated, even to the point of death? The coronavirus is survivable, but for those who are already compromised with pre-existing health conditions, it can kill just like the flu, yet reports, as we shall see, immediately began to flood into American mainstream media and social media of mass deaths everywhere the virus showed up. Conspiracy theorists point to the many-pronged attacks on our health from things like chemtrails, toxins in food and water, pollution, geoengineering, genetically modified organisms (GMOs), drugs and pharmaceuticals, EMR and DEWs, vaccines, and now the possibility that a virus many believe to have been engineered in a lab near Wuhan could interact with another source such as 5G and become far more deadly.

 Conspiracy theorists point to … the possibility that a virus many believe to have been engineered in a lab near Wuhan could interact with another source such as 5G and become far more deadly.

5G is going to be everywhere. It's only a matter of time. If the thousands of doctors, researchers, and concerned scientists who have warned world governments, the UN, the WHO, and the CDC of the potential dangers and hazards to our health are worried, shouldn't we be worried, too? Anything that compromises our immune systems and our inherent ability to fight off disease is something we should all be speaking out against. Aren't we sick and tired enough already?

Environmental Pollution Revisited

Progress often means an increase in environmental pollution, and now we have 5G technology to add to the existing problems. The Industrial Revolution of the nineteenth century saw an explosive increase in pollution of our skies, as pollution itself occurs when nature cannot destroy or eliminate foreign substances and elements without damaging itself in the process. So now we have several types of pollution to battle:

Industrial—Thanks to the development of fossil fuels, industrial pollution has been with us for quite a while, especially from the use of coal to make machines work faster and more efficiently and the rise of smokestacks pouring pollutants into our air and factories pouring sludge into our waterways. The use of coal even replaced many human jobs because it was cheaper to use. The building of dams, power plants, and electricity plants increased, as did the demand. Growing businesses demand growing amounts of heat, electricity, and water, and this results in increased amounts of waste.

Transportation—Horses and buggies left no more pollution on roads than a pile of manure. Today, we have vehicles of every size and shape on the ground, in our waterways, and in the air giving off their toxic pollutants. Our freeways and highways are clogged messes of exhaust fumes and gases. Trains were once propelled by massive amounts of coal. Planes dot our skies in increasing numbers, spewing their exhaust into the atmosphere.

Agriculture—Our vast farmlands often come at a price in the destruction of the original environment and the use of tons of pesticides that poison the soil and the air from spraying. The vast amount of water needed to keep farm crops alive is overwhelming, and for some products or foods, such as beef, huge sections of green forests are cut down to accommodate grazing cattle, contributing to climate change.

Housing—At home, we also create tons of pollution via our garbage, water waste, fireplaces and charcoal grills, and the many toxins found in the paint, wood floors,

Protesters in London, England, express their fears of the dangers of 5G near Parliament.

carpeting, and other items used to build our homes, not to mention the toxic cleaning agents we keep in our garages.

Pollution and toxins from every direction threaten our ecosystem, and individuals and environmental groups try but often fall far short of improving the situation on anything other than a local basis. We did see a reduction in the ozone layer, thanks to a concerted global effort, but as we get ready to pollute our skies with the latest and greatest technologies and alter the very fabric of our cells and DNA in the process, one wonders if the ecosystem can continue to hold. It is only as strong as its weakest link, and we humans are doing a great job of weakening almost every link.

Aside from any conspiracies to cut down the population by dumping toxins in the air or using microwave radiation to cause fertility rates and sperm counts to bottom out, the sheer lust for an easy way of life contributes to much of the pollution we deal with daily. Changing our habits is hard, and perhaps it's easier to blame someone else for our inability to do so. If corporations and government agencies don't care about our health—and that has been quite evident looking at the pollution and the concerted efforts to deny any problems associated with it—doesn't it behoove us as citizens of this country to take control and demand they start giving a damn?

At the writing of this book, more countries have decided to join the ban of 5G, along with many small towns and larger cities across the United States. Even this author's hometown is protesting the rollout here. It will take a lot of raised voices and maybe even a boycott of cell phones and gadgets to force the hands of the industry, and it raises the question of just how much we put our health as a top priority above convenience, fun, ease, and our addiction to constant communication with our cell phones and other toys. Can we do it?

During the 2020 Super Bowl, Verizon Wireless debuted its 5G commercials and began advertising all over social networking. In March 2020, AT&T began touting a new technology called 5G AirGig, a groundbreaking way to turn existing power lines into transmitters. Its press release read: "We hope one day there will be no need to build new towers or bury new cables in locations close to aerial power lines. Instead, using AirGig patented technology, we would install devices to provide high speed broadband which can be clamped on by trained electrical workers in just a few minutes."

 What AT&T never explained about its AirGig devices is that they will make it virtually impossible for anyone to escape exposure from wireless radiation.

What AT&T never explained about its AirGig devices is that they will make it virtually impossible for anyone to escape exposure from wireless radiation. Dafna Tachover, director of the Stop 5G and Wireless Harms Project at the Children's Health Defense, a nonprofit organization led by Robert Kennedy Jr., wrote in "5G AirGig: What is It and Should You Be Worried?" on March 17, 2020, that "AirGig will saturate our environment—every inch of it—with close-proximity, high-intensity radiation. The few relatively safe areas that still exist will quickly disappear." There will be nowhere to escape to from the negative health effects of wireless radiation exposure now being rolled out in cities all over the country. "At present," the article continues, "the deployment of 5G promises to interconnect 20 billion devices wirelessly, adding 800,000 'small cells' (base stations) close to our homes and launching 50,000 satellites that will also require 1,000,000 antennas on the ground."

It's exciting to think that you can call and text faster, gain access to the internet quicker and easier, and not wait for movies on streaming services to buffer on your gadgets. But amid the excitement of new generations of existing technology are threats to our minds and bodies, and those of our children, threats that have warranted the outcry of hundreds of doctors, scientists, researchers, and health advocates. At the very least, we should balance our love of technology with a love of our own well-being.

WHAT'S IN THE WATER?

We can survive only a few moments without air. We can survive only about three days without drinkable water. Despite living on a planet with huge oceans, lakes, and rivers, access to drinkable, clean water diminishes each day. The sheer amount of toxins in the water we drink is mind-boggling, yet water is life to the human body, which itself is made up of approximately 60 percent water. Our brains and hearts are made up of 73 percent water, our skin is 64 percent water, our blood is 83 percent water, our liver is 86 percent water, our lungs are about 83 percent water, our bones are about 31 percent water, and our muscles and kidneys are 79 percent water. Imagine how our bodies function when the water we take in for hydration and replenishment is filled with chemicals, toxic waste, sludge, factory runoff, and a huge variety of hormones and pharmaceuticals that all influence our health and behavior.

Then we buy water in plastic bottles, thinking it cleaner and safer, only to learn it is not only just glorified tap water, but the bottles themselves create new problems. The plastic leaches toxins into the water (more on that in a bit), and the bottles themselves become a major pollutant in our landfills and oceans.

Water pollution increased at the same time air pollution increased, thanks to growing industrialization and the waste produced by factories, factory farms, and manufacturing plants. Now, the toxins that were being spewed out of smokestacks were also flowing out of large pipes into our creeks, rivers, lakes, and oceans. Meanwhile, over the decades, drinkable water from snow and mountain runoff was vanishing thanks to climate change, those who wished to privatize water were making inroads to do so, corporations such as Nestlé and Coca-Cola were diverting precious water resources from poor countries and farmers for their monster factories and bottling plants, more cattle were grazing on land for meat products and requiring more drinking water, and our population was increasing a billionfold, meaning more people who need water.

As progress advanced, our use of water increased, along with our wastefulness of it, but sadly, so did the pollution and toxins we dumped into it. According to the National Institute of Environmental Health Sciences, over two billion people drink contaminated water that is harmful to their health. As of 2018, there were about 250 different carcinogens in the water systems, and with the increasing amounts of toxins, chemicals, pesticides, dyes, and pharmaceuticals, no doubt that number will increase each year.

 According to the National Institute of Environmental Health Sciences, over two billion people drink contaminated water that is harmful to their health.

Water is considered a universal solvent, able to dissolve more substances than any other liquid on the planet, but that's what makes it so susceptible to pollutants and toxins. All those poisons that come from adjacent farms, factories, manufacturing plants, and other polluting sources continuously fill our waterways, not to mention the more highly visible human and animal waste, trash, oil from spills, fertilizers, and microorganisms. This harms not just humans but also life found in our ponds, lakes, rivers, streams, and oceans.

According to the Natural Resources Defense Council (NRDC), the polluting of groundwater from contaminants such as pesticides and fertilizers is a huge problem because approximately 40 percent of Americans get their water from these underground storehouses. These aquifers are often rendered unsafe for consumption thanks to wastewater leached from nearby landfills and septic systems.

The planet is covered by approximately 70 percent surface water in the form of oceans, seas, lakes, rivers, and so on. More than 60 percent of the water we drink and use in our homes in the United States comes from surface water sources other than the ocean, and the Environmental Protection Agency (EPA) has determined that roughly half of our rivers and streams, and more than one-third of our lakes, are unfit for drinking, fishing, and swimming because of various pollutants. Excessive amounts of nitrates and phosphates from farm waste and runoff from fertilized crops are the major pollutants found in water as well as municipal and industrial waste and the pollution that individuals dump into our waterways or flush down toilets.

Our oceans are particularly affected, with over 80 percent of the pollution found there originating from sources onshore, such as factories, plants, and farms. In addition, we have the occasional oil spill and leak, the absorption of carbon pollution from the air into the ocean waters, stormwater runoff and debris, and illegal dumping.

We treat drinkable water as if it is an infinite supply that will last forever no matter how much we waste. First of all, it is not an infinite supply. Secondly, it is barely drinkable. The NRDC identifies several types of water pollution beyond groundwater, surface water, and ocean water:

▲ Point source—Contamination that originates from a single source, such as wastewater, oil refinery spills, illegal dumping, manufacturing runoff, or septic systems.

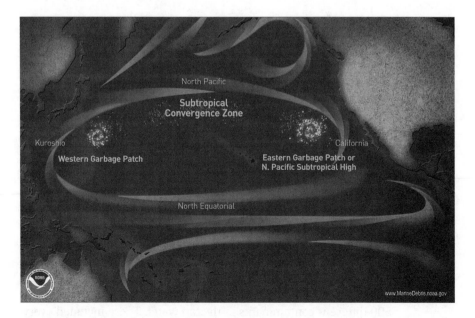

Today, there is so much garbage floating in our oceans that giant garbage patches have formed, some of which are as large as the state of Texas!

▲ Nonpoint source—Contamination from diffuse sources such as agricultural or storm runoff or debris that is blown into waterways. This is the leading cause of water pollution in the United States.

▲ Transboundary—Contaminated water from one country that spills into another country. San Diego, California, struggles with wastewater from Tijuana, Mexico. This can also occur after a major disaster or oil/chemical spill.

▲ Agricultural—Livestock and farming waste and runoff, pesticides, bacteria and viruses, algae blooms, and more come from the largest consumer of fresh water in the world and one of the biggest polluters.

▲ Wastewater—Contaminants from used water from toilets, showers, sinks, sewage, and stormwater runoff. More than 80 percent of this wastewater goes right back into the environment untreated. The United States does process about thirty-four billion gallons of wastewater per day back through water treatment facilities, but it releases more than 850 billion gallons untreated into the environment every year.

▲ Oil—Contamination from major and minor oil spills, leaks, land-based factories and farms, ocean tankers, the shipping industry, and leaks from cars and trucks. About 1,000,000 tons of oil enter the environment each year from these and other sources.

▲ Radioactive substances—Contaminants from the production and testing of military weapons, radioactive medical waste from hospitals and labs, and leaks and accidents at nuclear facilities. All of this creates waste that remains in the environment for thousands of years.

Tap Water

We rely on tap water. We drink it, bathe in it, and wash our clothes, dishes, and cars with it. Yet the Environmental Working Group (EWG) performed studies from 2010 to 2015 on over 50,000 different water utility companies in the United States and found over 500 different contaminants in the tap water. These included everything from pharmaceuticals to pesticides to bacterial agents to fluoride to disinfection products to lead from old pipes and a whole lot more, all of it toxic to the human body. They reported that the entire nation's water supply was "under assault from a toxic slew of pollutants." Some of it originates at the source of our groundwater, and others are introduced in the treatment process and the delivery via pipes to our sinks and faucets.

Farming contributes to poisoned tap water by increasing the number of algae blooms from manure and sewage runoff as well as the herbicide used to treat the algae blooms. These herbicides contain glyphosate, which is carcinogenic and is the active ingredient in Roundup, which is being banned in many states because of its toxic effects on the human body. Concentrated animal feeding operations at factory farms contribute nitrates into the water systems, which can cause cancer, childhood diabetes, and birth defects.

Fertilizer runoff from farms gets into lakes and oceans, causing algae blooms that produce toxins that are dangerous to wildlife and humans.

Our tap water also includes per- and polyfluoroalkyl substances (PFAS), which are also found in consumer products such as cookware and stain repellents. These stay in the human body for long periods of time and have been linked to cancerous tumors and reproductive developmental problems. Once you turn on the tap, you are also consuming contaminants that are harmful to your health, such as volatile organic compounds (VOCs), acids, mercury, and radioactive contaminants that are not fully excreted during water treatment processes.

Dirty Dairy

Massive dairy farms, called concentrated animal feeding operations (CAFOs), can be so big they run the length of six or seven football fields and contain over 15,000 cows. The animal abuse is abhorrent, as is the amount of water it takes to operate a farm of this size. One cow alone consumes over 150 liters of water per day. But that's not all; these CAFOs use incredible amounts of water to operate, not just for the animals to consume but for cleaning the facilities, agricultural runoff, and illegal waste disposal.

This adds up to contaminated aquifers and waterways anywhere near a dairy operation. Poor storage of waste can lead to massive manure spills, and these have been cited in states like Michigan, Minnesota, and Washington. In 2013, 15,000,000 gallons of manure and agricultural waste spilled into a slough that drained into Washington's Snohomish River.

Mining

According to a February 2019 Associated Press story, "50 Million Gallons of Toxic Water Flow into Lakes and Streams Every Day in the US," just one industry alone is responsible for placing tens of millions of gallons of contaminants such as arsenic, lead, and dangerous metals into lakes, streams, and other drinking water *sources every day:* the mining industry.

A damning report exposed companies mining for gold, silver, lead, and other minerals that were dumping toxins and contaminants into water sources in California, Colorado, Montana, Oklahoma, and other states and deliberately turning the cleanup bill over to taxpayers. The AP investigation looked at public data and

Mining operations of all kinds result in toxic metals and other pollutants, such as lead, arsenic, and mercury, being leached into rivers and streams, which eventually end up in our drinking water.

research on forty-three specific mining sites that were under federal oversight yet were responsible for 50,000,000-plus gallons of contaminated wastewater daily finding its way into ponds, rivers, soil, groundwater, and other sources. For the water to be treated, the cost is astronomical, and these corporations found ways of making sure they didn't pay for it. Many of these mining sites are called Superfund cleanup sites, which are supposed to be the most hazardous according to the EPA, but little is ever done to abate the pollution, thanks to cutbacks and lack of oversight.

Forever Chemicals

GMO Free USA reported in December 2019 about research from the National Atmospheric Deposition Program, working with the U.S. Geological Survey (USGS), that rainwater in various parts of the United States had high levels of toxins called "forever chemicals" because once in the environment, they last forever and do not degrade. Exposure to these chemicals is linked to cancers in humans and animals, most notably cancer of the kidneys and the testicles.

There are more than 4,700 variants of forever chemicals labeled toxic, yet the EPA only regulates two of them. Harmful chemicals are commonplace in food packaging, insulation, carpeting, cookware, and firefighting foam. These toxins appear to enter the atmosphere via factory emissions and then fall to the ground as rainwater. High concentrations of PFAS, for example, have been found in samples taken in rainwater near factories and chemical plants in North Carolina, and high levels found near military bases because of firefighting foams that make their way into the atmosphere have resulted in major health issues. Yet these chemicals are left off the EPA's Safe Drinking Water Act's list of banned contaminants.

Sludge

Just the word *sludge* sounds awful. Sludge is a semisolid slurry produced by industrial processes such as water treatment, sanitation, and wastewater treatment. It is also referred to as sewage sludge and has a soupy texture filled with biological and chemical products. Food-processing plants create their own protein-heavy sludge that is beneficial and can be used for animal feed and isn't disposed of in landfills like ordinary sludge.

Sludge that comes from on-site sanitation systems such as septic tanks is called "septage" and, as fresh sewage, is fairly solid, like a hardened mud. Sewage sludge is treated to reduce its volume by decomposing it by exposing it to anaerobic bacteria in treatment plants. Biosolids are the result of the sewage sludge being treated for reuse.

Each year, millions of tons of sewage sludge is produced, often filled with bacteria and pathogens, micropollutants, heavy metals, lead, arsenic, cadmium, chemicals, hormones, barium, and other hazardous substances that have limits defined by the EPA but often surpass the ceiling with little punishment. Some of the toxins were banned, such as PCBs in 1979, but they were still found in many plants in South Carolina in 2013. No matter the laws and regulations, there always seems to be a way around them, and our water sources pay the price. Other contaminants come from human-use products such as personal care toiletries, medicines, and tissue and toilet paper.

No matter the laws and regulations, there always seems to be a way around them, and our water sources pay the price.

Once treated, depending on the production process, sludge either ends up in landfills or is dumped in our oceans or applied to lands as fertilizer. In 1988 President Ronald Reagan signed a law prohibiting ocean dumping, but somehow sludge ends up in our open waters due to lack of oversight and poor regulations. There are international and domestic laws in place banning ocean dumping, yet our oceans continue to be filled with sludge, wastewater, and tons of trash, such as plastic bottles, killing marine life and choking off the food chain due to toxins.

Pharmaceuticals

One of the most problematic forms of water pollution comes from our proclivity toward taking pills and medicines. Once our bodies have digested pharmaceuticals, the rest ends up as waste, entering our water systems once we flush the toilet. Even though levels may be in the parts per trillion, added up over time and with a rising population taking medications, you have a long-term exposure to a cocktail of drugs and hormones that could cause harm and disrupt physiological processes.

Think of the millions of people in the United States alone who take at least one pill for some ailment or illness. Many take over five pills a day. Even if a tiny percentage of that enters the water system, that means you end up drinking it or bathing in it over time enough to have some effect on your health. These drugs include birth control pills, hormones, antipsychotics, antidepressants, anxiety meds, aspirin and other painkillers, antibiotics, and endocrine disruptors like steroids, which have been shown in studies of fish to alter hormones, decrease fertility, and cause sex changes. One study by Sébastien Sauvé, associate professor of environmental chemistry at the Université de Montréal, published in *Chemosphere,* showed three different antidepressants in the livers of brook trout exposed to municipal wastewater and runoff.

Sauvé suggested that these drugs have a cumulative effect. Antibiotics particularly could create antibiotic resistance if they accumulate enough in the water and are then consumed by enough people. Then there are the more vulnerable populations of the elderly, infants, and those with immune-compromised systems consuming drugs in even micro doses that might impact their health. There are no guidelines or thresholds, Sauvé warns.

A 2010 study at Trent University in Peterborough, Ontario, Canada, showed the presence of six different antidepressants downstream from wastewater treatment plants in southern Ontario. We must ask what these drugs are doing to fish and other aquatic life, not just humans. Humans consume some of those fish, which means we are also getting exposed to chemicals and drugs via our food sources.

According to the June 2011 article "Drugs in the Water," published by Harvard Medical School's *Harvard Health*, our bodies only metabolize a fraction of the drugs we take. "Most of the remainder is excreted in urine or feces (sometimes sweated out) and therefore gets into wastewater. An increasing number of medications are applied as creams or lotions, and the unabsorbed portions of these medications can contribute to the pollution problem when they get washed off."

This doesn't include the millions of unused pills people flush down the toilet rather than properly disposing them or the liquid medicines people take for colds, allergies, and stomach issues. We also need to consider the creams, lotions, shampoos, and other liquid products used every day by millions of people. It all ends up somewhere and, unfortunately, that's our wastewater.

Obviously, one way to stop this from being too harmful is to have better regulations for treating wastewater and more test-

How to Properly Dispose of Unused Pills

In order to reduce the "pharmaceutical footprint" as a result of drugs entering the water system and soil, here are some suggestions on what to do with those unused bottles of pills and meds:

▲ Participate in a drug take-back program in your community. Ask your pharmacist, as many pharmacies are also offering take-back programs.

▲ Many pharmacies offer bins into which you can throw your medications, and they will properly dispose of them for you.

▲ *Never* flush pills or meds down the toilet or sink.

▲ If you choose to throw meds in the regular trash, put them in a plastic bag filled with cat litter, sand, water, or coffee grounds to cut down on the chances a child or animal might eat the contents. Remove identifying labels from plastic bottles before putting them in the recycling bin.

▲ Do not do bulk pill orders unless you are certain you need to take them all.

ing of water to assure chemicals and toxins are well below acceptable levels. It's also the drug manufacturers polluting the waterways. A 2011 USGS study found that contamination levels downstream of two major drug manufacturing plants in New York state, according to *Harvard Health*, were 10 to 1,000 times higher than other facilities around the country. "Sewage plants are not currently designed to remove pharmaceuticals from water. Nor are the facilities that treat water to make it drinkable," *Harvard Health* reports. This is terrifying. Though some parts of the sewage treatment process do remove pharmaceuticals from the initial water, it ends up in sludge, which may be used as fertilizer, which means you are getting some extra pharmaceuticals in your food grown with that fertilizer.

Hormone Disrupters

Pharmaceuticals in the waterways, notably estrogen, are contributing to a rise in intersex fish with both male and female sex organs and characteristics. Fish living downstream of wastewater treatment plants have been linked to increased female and intersex characteristics because of the levels of estrogen. One has to wonder how estrogen in the environment might affect human beings.

Professor Charles Tyler of the University of Exeter in England took part in a 2008 study that found almost one-quarter of male roach fish taken from fifty-one different sites along English rivers showed signs of feminization,

You should never just throw away medications or flush them down the toilet. You don't want those chemicals getting into the water supply.

including having eggs in their testicles. Many showed high levels of estrogen, which is used in birth control pills and other hormonal drugs. Tyler told the British news source *The Independent* in July 2017 that of the thousands of chemicals used by humans that have an effect on the natural world, at least 200 result in feminization and that this can result in difficulty breeding. "If they are moderately to severely feminized, they are compromised as individuals and they really struggle to pass their genes on." Over time, this translates to a reduction in population. "If we get sufficient evidence indicating there's a high likelihood of a population effect, perhaps we need to be more proactive about restricting—or banning—these chemicals."

Could humans be experiencing the same effects?

⚠ Testosterone levels in men are affected by endocrine-disrupting chemicals (EDCs) to the point where levels are beginning to decline earlier and earlier, sometimes even in childhood.

Testosterone levels in men are affected by endocrine-disrupting chemicals (EDCs) to the point where levels are beginning to decline earlier and earlier, sometimes even in childhood. The normal man will see levels decline from age thirty on, but because of chemical exposure to EDCs, and the escalating amounts released into the environment, both animals and humans are seeing a disruption to endocrine systems that some describe as having "gender-bending propensities." In "Testosterone Decline: How to Address This Challenge," Dr. Joseph Mercola examines research indicating that plastic products such as containers and bottles are part of the threat, along with wastewater, as we've seen earlier.

Pediatric studies show both girls and boys are developing sexually at an earlier age, with girls getting their first periods and developing breasts years earlier, sometimes at the age of nine or ten, and boys developing large testicles, pubic hair, and facial hair by the age of nine. Mercola writes that this early puberty "can significantly influence physical and psychological health, including an increased risk of hormone-related cancers." EDCs have been shown in numerous studies to cause testicular dysgenesis syndrome (TDS), which threatens male fertility and increases the risk of testicular cancer.

One of the worst EDCs is phthalate—a chemical often added to plastics that can, Mercola states, "have a disastrous effect on male hormones and reproductive health." Phthalates are also linked to birth defects in male infants and reproductive issues in women, including infertility, premature ovarian failure, and nonreproductive disorders. Sadly, they are found in many common products, such as soaps, shampoos, deodorants, perfumes, plastic bags, food packaging, and vinyl flooring.

Other EDCs include perfluorooctanoic acid (PFOA), a carcinogen used in water and grease-resistant food coatings; methoxychlor, an insecticide that a Japanese study found impaired mating behaviors in quails; nonylphenol ethoxylates (NPEs), powerful endocrine disruptors that mess with your gene expression; and the most well known of all, bisphenol A (BPAs), which are found in some plastic bottles, plastic products, food cans, and sealants. They've been found to alter fetal development and have also been linked to breast cancer.

Because the endocrine system is made of hormones and glands that regulate every bodily function, disrupting them via chemicals and contaminants is not something to be taken lightly. These hormones play a critical role in every aspect of our physiological development, metabolism, and behavior. Pesticides often contain chemicals that alter our hormonal system, as do plastics, birth control pills, and a host of other products we will examine in a later chapter.

There are well-documented studies that show high doses of endocrine-disrupting chemicals and contaminants cause gender-related effects, but for lower doses, it's harder for many scientists and researchers to find the threshold of concern. However, during crucial times of development, such as intrauterine exposure and infant exposure, even low doses are a huge concern. Exposure during these critical windows of development could have permanent effects on overall fetal development and functioning of the sex organs and endocrinal system as an adult.

Past research links EDC exposure to reproductive abnormalities in birds, frogs, fish, alligators, seals, polar bears, and other wildlife. Lake Apopka in Florida, for example, is polluted by farming wastewater, which has led to feminized alligators. Males have shorter penises and lower testosterone levels.

Many plastic bottles are now BPA-free, but you should carefully examine those with a recycle code of 3 or 7 on them because some of these still may contain bisphenol A.. Those marked 1, 2, 4, or 5 are safe, as well as many plastic bags.

Why, then, would we think humans are immune to this? Many people even go so far as to suggest that the presence of these hormone and endocrine disruptors in our water and products is responsible for the growing levels of gender dysphoria in today's children, teens, and young adults. Scientists are quick to point out that this kind of disruption is more of a threat to entire populations because of lower fertility rates and decreasing populations; some wonder if the hormonal shifts and effects are part of a cause for concern on an individual level.

The issue of gender dysphoria is fairly new and utterly controversial when it comes to human beings. Different camps suggest a variety of reasons why young people who are one gender may identify as another: behavioral issues, psychological issues, genetic issues, pressure from peers, and anger toward parents are a few. But the increasing presence of chemicals that confuse the body as to its own gender identity might also confuse the mind and behavior of that entity; it's an area that deserves more study and examination. This is especially critical if the disruption decreases fertility rates of both males and females to the point that it affects population levels. Any chemical that mimics or disrupts sex hormones, such as BPAs, can also cause other issues aside from gender disruptions, such as prostate and breast cancer, aggression, depression, and even suicide because of the associated pressure from peers and family, guilt, and shame over gender confusion.

Plastic Poison

To avoid the chemicals and contaminants we are told are in our tap water, we turn to bottled water instead. We assume it's cleaner and less toxic than whatever is in our tap water. Yet the same EDCs are found in plastics, such as plastic water bottles, and leach into the water we drink on a daily basis. The biggest offenders, again, are BPAs, which are used to make the same plastic bottles. BPAs mimic estrogen, the female sex hormone, which in turn lowers the sperm count in men. BPAs are also found in the dye used on store sales receipts and parking tickets.

 BPAs mimic estrogen, the female sex hormone, which in turn lowers the sperm count in men.

Ecotoxicology professor Tamara Galloway of the University of Exeter studied the blood and urine of ninety-four teenagers between the ages of seventeen and nineteen. The study found that 80 percent of the teens had EDCs in their bodies. Interestingly, when the same teens changed their diets to include more fresh foods and produce, the BPA levels began to fall among those with the highest initial levels. The others were not as affected.

While scientists and researchers put out studies showing the dangers of these chemicals, corporations put out their own research studies showing there are no dangers at all. With BPAs, and all the studies showing they disrupt a basic human system, you have numerous food lobby groups, plastic companies, and industry groups, such as the British Plastics Federation (BPF) and the Food Standards Agency (FSA), claiming, as the BPF did in regard to the above study, "At current exposure levels, plastics containing BPA pose no consumer health risks." According to the FSA, "Minute amounts of BPA can transfer from packaging into food ... but independent experts have advised that these levels of exposure are not considered to be harmful."

Who do you think is telling the truth? The researchers who find these chemicals are causing harm to human health yet have no external motivation or industry association, or the lobby groups, industry representatives, and "independent experts" hired by the polluters who do? Pro-industry research studies are rampant when it comes to environmental poisoning of our air, water, and bodies, so always follow the source to the money and the money to the source.

The EPA admits tap water is often contaminated with small amounts of minerals, nitrates, microorganisms, and toxins that can infiltrate the water system but adds that bottled water is often just glorified tap water and has the added problem of plastic container chemicals. You might be better off with tap water, depending on the water quality in your area, which you can learn by contacting your local water department. Also keep in mind the testing of bottled water is not mandatory, and you have additional issues associated with what happens at the bottling plant. Bottled water is also more expensive to produce, which means more energy wasted, and the bottles result in tons of trash that can't always be recycled, meaning it ends up in our landfills and, worse yet, our oceans.

It's a crapshoot. While the EPA does test tap water, it is often compromised after a major storm or from old pipes, crumbling infrastructure, and ancient delivery systems despite the signing of the Safe Water Act in 1974 that authorized the EPA to set national standards for our drinking water and protect us from both natural and man-made contaminants. Federal law dictates that when water is not safe to drink, the public is to be informed. That may be the case, but often neighborhoods go without drinkable water for months, even years.

Flint, Michigan, Case Study

The water crisis in Flint, Michigan, began in April 2014 when the City of Flint replaced its water source from the treated water of the Detroit Water and Sewage Department, which got its water from Lake Huron and the Detroit River and promptly treated it with orthophosphate to coat the pipes and prevent leaching of lead, to the cheaper Flint River, whose water went through decades-old pipes. During the switch, those in charge did not apply corrosion inhibitors, and the aged pipes leaked lead into the water supply, causing a yellow-brown water to come from the faucets of residents that not only tasted bad but also smelled bad. The levels of lead were high, exposing over 100,000 people who lived in the area to the neurotoxin, and forcing Michigan governor Rick Snyder to declare a state of emergency.

Then President Barack Obama stepped in by announcing a federal state of emergency, allowing the Federal Emergency Management Agency (FEMA) to assist in the situation, but by then over 12,000 children had already been exposed to the lead. Lead was only one problem. The water source change also led to an outbreak of Legionnaire's disease, which killed twelve people. A February

National Guard troops help with the distribution of bottled water during the crisis in Flint, Michigan, in which the local government changed the city's water source to the Flint River without treating it properly.

2019 test of water systems in Flint—even those homes with the worst water—showed that water was safe to drink now throughout the city. After $400 million in funds was received from the federal government, the city now has replaced its water lines, secured a safe source of water, and provided homes with water filters. Even so, most residents refuse to drink anything but bottled water because they no longer trust what authorities tell them.

Lead is considered a "forever chemical" that, once in the body, stays there for the rest of your life. Forever chemicals are linked to cancer and other diseases and due to aging pipelines, Michigan is considered one of the most polluted states when it comes to water supply. Flint is also dealing with the issue of corporations such as Nestlé, which struck a deal with the city in 2018 to pump water at the rate of 250 to 400 gallons per minute, taking all the clean water for its needs, leaving the residents with the choice of paying for expensive bottled water or taking their chances on what comes out of their taps. Luckily, legal action on behalf of the citizens forced the state's second-highest court to side with citizens and ruled that Nestlé's bottling operation was not an essential public service. To this day, however, Nestlé continues to attempt to open a bottling operation in the small town of Osceola Township.

Tap water is also treated via chlorination and may have a higher mineral content, so it's important to know what is in your local water, especially if you have an immune-compromised system, cancer, heart disease, or are using the water in infant formula. Water departments usually send mailers to residents with water quality reports, or you can request one by phone or on their website. Know what is in your tap water, but be aware that bottled water is filled with toxins, too.

 Most bottled water is filtered tap water, but it contains a number of contaminants to be aware of.

Don't be fooled by commercials promoting mountain springs and natural lakes and water sources. Do your research to find brands that are consistent with their own claims. Most bottled water is filtered tap water, but it contains a number of contaminants to be aware of. Researchers at the University of Iowa's Hygienic Laboratory conducted comprehensive tests on ten of the leading bottled water brands. They got the water from retailers in nine different states. They found a total of thirty-eight toxic pollutants, and each brand had an average of eight of them. This includes fluoride, caffeine, pharmaceutical drugs, fertilizer residue, plasticizers, solvents, fuel propellants, radioactive isotopes, and arsenic. Four of the brands tested proved positive for bacterial agents.

Sadly, over a third of the above chemicals are not regulated by the bottled water industry, which follows voluntary standards, if any. One of the study's authors concluded that "this bottled water was chemically indistinguishable from tap water." The study author pointed to the media campaigns that make this bottled water appear to come from natural glaciers and mountain springs. One brand of water was applied in a study at the University of Missouri to breast cancer cells and resulted in a 78 percent increase in cancer cell proliferation rates when compared to untreated cells! This is the result of estrogen-mimicking contaminants that are in the water from pharmaceuticals and birth control pills, from the leaching of the plastic bottles, or both.

According to the Environmental Working Group (EWG) in the October 13, 2008, article "Harmful Chemicals Found in Bottled Water," the purity of most bottled waters cannot be trusted. Over the last ten years, strides have been made by bottled water companies to try to clean up their act, but even the most recent studies show continued inclusion of contaminants and plastic particles. The

EWG writes, "Unlike tap water, where consumers are provided with test results every year, the bottled water industry does not disclose the results of any contaminant testing that it conducts." The industry claims its water is as safe as, if not safer than, tap water, but until all the studies they conduct are transparent to the public so that we can not only see the results but also compare with independent nonindustry studies, it really comes down to the consumer doing a bit of legwork to find the best bottled water source.

Microplastics

Microplastics are particles of plastic found in bottled water. So, in addition to chemicals, be aware of the tiny little particles of plastic you may be consuming on a regular basis. In the largest study of its kind, a journalism group called Orb Media examined 250 bottles collected from nine different countries. They found an average of ten plastic particles per liter. These particles are larger than the width of one human hair. Sherri Mason, a chemistry pro-

Numerous brands of bottled water available in American grocery stores were tested and found to contain microplastics.

fessor at State University of New York at Fredonia, conducted the mass study. She analyzed the brands and found plastic in "bottle after bottle, brand after brand." Though she claimed there were minimal if any studies showing the harm of ingesting microplastics, she claims the results were concerning, nonetheless. Obviously, new studies looking at BPAs refute that claim. Plastic leached into water and consumed by humans is harmful. We know that now.

The water brands in the study included Aquafina, Dasani, Evian, Nestlé Pure Life, S. Pellegrino, Aqua, Bisleri, Epura, Gerolsteiner, Minalba, and Wahaha.

Another large study commissioned by the Story of Stuff Project looked at nineteen different bottled water brands. Among them were Aquafina, Arrowhead, Boxed Water, Crystal Geyser, Dasani, Deer Park, Eternal Water, Evian, Fiji, Glacéau Smart, Ice Mountain, Icelandic Glacial, Ozarka, Penta, Poland Spring, Texas Spring Water, Trader Joe's Mountain Spring, True Zealand, and Zephyrhills. Abigail Barrows of Ocean Analytics headed the study, which sampled each brand under a microscope for microplastic contamination.

The most highly contaminated brands she found were Boxed Water, Fiji, Ozarka, Evian, Icelandic Glacial, and Crystal Geyser. Barrows concluded that "people are directly ingesting plastic particles when drinking most types of bottled water."

The WHO's Jennifer De France told reporters in 2018, according to *Business Insider*, that "while we think that the risk to human health is low, this is based on limited evidence and we recognize there is a need for more research." Just because data in the Orb Media study didn't point to human health issues from consuming microplastics does not mean they are not doing harm. Bruce Gordon of the WHO told the same reporters that consumers shouldn't worry, but he suggested concerned citizens focus on drinking tap water to "reduce pollution."

 There are studies showing that this type of synthetic clothing is found in the bellies of fish and sea life.

By the way, the majority of plastic particles in bottled water are polypropylene from the cap and polyester and polyethylene terephthalate from the bottles themselves. Do you want to consume any quantities of these chemicals? If not, check into the safety of your tap water or buy a high-quality water filter, preferably a carbon filter or a more

Water Filter Choices

When looking for a home water filter, there are three main types to consider.

The most affordable and common are granular carbon and carbon block filters, which remove organic and industrial contaminants, including pesticides, herbicides, and chemical agents. These come in countertop models and under-the-counter filters attached to the pipes.

Reverse osmosis (RO) filters are more costly and need to be cleaned often but remove more of the organic and inorganic contaminants, including fluoride. These also may require professional installation to be sure they are operating properly. The downside is they also remove beneficial minerals from tap water.

Ion exchange filters soften water and remove dissolved salts and are fairly easy to maintain but are more subject to contamination from organic matter, chlorine, iron, and bacteria.

It's a good idea to contact a water filter company and tell them your needs or ask for assistance at a home store. If you are looking just to filter your drinking water, you may not need something complicated. Some people use filters in the shower and bathtub or prefer to have a water filter attached to their piping system and not worry about countertop clutter and replacing water pitcher filters so often. Match the filter to your needs.

expensive reverse-osmosis filter system if you live in an area known for its less-than-stellar tap water, and drink your tap water with even less concern for toxins getting into your system. And whatever you do, follow the source to the money and the money to the source when considering studies and who or what entity is behind them.

On a side note, according to the International Union for Conservation of Nature, one of the main culprits for *how* these microplastics work their way into bottled water outside of the plastic bottles themselves comes from the washing of synthetic clothing that is discharged via washing machine waste. There are studies showing that this type of synthetic clothing is found in the bellies of fish and sea life. Even the washing machine is contributing to the pollution of our water sources, and this doesn't include the chemicals found in detergents and fabric softeners.

Don't be fooled either by gorgeous images of flowing streams or the words "bottled at the source." Chances are very good that the source in question is a filthy bottling plant that happens to be within a mile of a pretty stream. The source is not a crystal-clear lake or mountain creek. It's where the product is bottled and labeled, and that can be anywhere.

Reduce Your Exposure

There are many ways to reduce your exposure to BPAs and other chemicals and toxins in plastic products.

▲ Carry and use glass, ceramic, and steel water bottles and fill with filtered tap water.

▲ Avoid canned formula for your baby and reduce canned foods in your home.

▲ Look for baby bottles that say "BPA-free."

▲ Do not touch the ink on store sales receipts.

▲ Use glass, porcelain, stainless steel pots and pans, and nonplastic containers to keep food in. Never cook food in a microwave in plastic containers.

▲ Keep plastic water bottles out of the car and away from the sun and heat.

▲ Look at plastic products with the number 7 recycling symbol on the bottom. If it does not also have a "PLA" or leaf symbol, it may contain BPAs. Look instead for plastics with recycling numbers 2, 4, and 5.

▲ Cut way back on buying bottled water not just for your health but for the health of the planet. Plastic bottles are filling our oceans and destroying marine ecosystems, despite recycling efforts.

Plastic Bottle Pollution

Plastic bottles are themselves a poison on our planet. Eighty percent of all plastic water bottles end up in landfills, which are already overflowing with millions of bottles. Every single bottle takes over 1,000 years to decompose, and while they do, they leak harmful toxins back into the environment. As of 2017, annual consumption was 1,000,000 water bottles *per minute*. They also make up a large percentage of all pollution found in our lakes, streams, and ponds and on our beaches.

Buying water in plastic bottles may be convenient, but it is destroying our environment despite the belief that these bottles can easily be recycled. In fact,

Plastic bottles manage to get in almost every nook and cranny of our land, oceans, and shorelines, and they take centuries to break down.

since 1950, the world has created over 6 trillion kilograms of plastic bottle waste and 91 percent of it has never been recycled, so it ends up sitting in our landfills or clogging our oceans, creating a growing menace to the planet. This is according to Jim Puckett, executive director of the Basel Action Network, who was interviewed for *Rolling Stone* magazine and stated, "When you drill down into plastics recycling, you realize it's a myth."

Mercola.com's March 18, 2020, report "Is Plastic Recycling Just a Big Fraud?" quoted *Rolling Stone* reporter Tim Dickinson as saying, "Unlike aluminum, which can be recycled again and again, plastic degrades in the course of reprocessing and is almost never recycled more than once." In some cases, recycling delays final disposal rather than avoiding it altogether. Sadly, plastic products can last for centuries wherever they are deposited.

According to *Environmental Health News*, "Two-thirds of all plastic ever produced remains in the environment," which helps explain why tap water, bottled water, sea salt, and a variety of seafood all come with a "side order" of microplastic. The article states that at this rate, "plastic will outweigh fish in our oceans by 2050. Already, plastic outweighs phytoplankton 6-to-1, and zooplankton 50-to-1, and more than half of the plastic currently inundating every corner of the globe was created in the last 18 years alone. With plastic pollution estimated to double in the next decade, it's quite clear we're traveling full speed ahead on an unsustainable path." Despite lawsuits against major plastic bottlers and distributors, the problem persists as more people shun tap water and buy bottled water.

This doesn't account for the tons of plastic bottles that end up in landfills, sewers, on the ground, in parks, campsites, and the like. New Mexico senator Tom Udall introduced the Break Free from Plastic Pollution Act of 2020, holding companies that profit from plastic accountable for pollution they create, but whether or not it is ever enforced is another question. Other legislation suggests banning single-use plastic items, especially when consumers themselves have failed to get on board with not buying them voluntarily.

The bottling industry, plastics industry, water companies, and global food and beverage companies also lobby hard against any regulations that can hurt their profits. The Mercola.com article points to a perfect example of the power of these lobby groups and their deceptive ad campaigns. Remember the "Keep America Beautiful" campaign back in the early 1970s? These were public service announcements asking people to stop littering by saying, "People start pollution. People can stop it." Few people realize this ad campaign was created by the polluters themselves as a clever way to shift blame from their production of plastic waste to the consumer. And even as they were shaming consumers as polluters, the industries behind the ad campaign were fighting bans on single-use packaging. That's the power of industry working with media to fool the end user, just like those fancy water bottles with deceptive pictures of gushing natural geysers and flowing mountain streams.

If consumers do not curb their buying of plastic water bottles, though, we will be facing an environmental catastrophe of epic proportions that our children and grandchildren will be stuck cleaning up, if it's not too late. Of course, continue to recycle everything you can, but it's not going to be enough. Here is what you can do:

As a U.S. senator from Arizona, Tom Udall sponsored the Break Free from Plastic Pollution Act in 2020. As of this writing, the bill is still being discussed at the committee level.

▲ Buy water in glass containers that can be washed and reused.

▲ Look for reusable plastic products such as shavers, cloth diapers, and grocery bags, and use old clothing for rags instead of paper towels.

▲ Buy a water filtration system, so you can drink your tap water safely.

▲ Buy glass food storage containers instead of plastic one-offs.

▲ Stop using plastic shopping bags and get cloth bags. Many stores give them for free or charge a quarter for them.

▲ For parties and picnics, use real cutlery and silverware instead of buying plastic ones that get thrown away.

By late 2020, consumption levels will have topped half a trillion bottles, and there is no way recycling efforts can keep pace. Many environmental groups believe plastic waste will rival climate change as a major crisis, especially as buying water in plastic bottles spreads to countries like China and India. The Ellen MacArthur Foundation estimates that by 2050, the ocean will contain more plastic by weight than fish. Right now, we pour over 13,000,000 tons of plastics into the world's oceans each year, which are broken down and ingested by birds, fish, and ocean creatures and organisms. Scientists at Ghent University in Belgium have proven that plastic has worked its way into the food chain in a study that calculated that people who eat seafood ingest up to 11,000 tiny pieces of plastic. A University of Plymouth study in 2016 showed that a third of fish caught in the United Kingdom contained plastic in their systems.

 The Ellen MacArthur Foundation estimates that by 2050, the ocean will contain more plastic by weight than fish.

Hugo Tagholm of the Surfers against Sewage marine conservation group told the *Guardian* in 2017 that "the plastic pollution crisis rivals the threat of climate change as it pollutes every natural system and an increasing number of organisms on planet Earth." He went on to say that current science proves plastics cannot assimilate into the food chain because of their toxins, and while the production of throwaway plastics such as water bottles has increased over the last two decades, "the systems to contain, control, reuse and recycle them just haven't kept pace."

Arsenic

According to the Centers for Disease Control and Prevention (CDC), arsenic enters the water supply from natural deposits in the earth or from industrial processes where it is used as an alloying agent, resulting in agricultural pollution created as a result of food preparation and irrigation of crops. "Naturally occurring arsenic" is a "metalloid" substance that is not a metal per se but shares qualities of a metal and dissolves out of rock formations when groundwater levels drop. Arsenic can be a particular problem in private wells. Miniscule amounts of arsenic are found in all rock, air, water, and soil.

Arsenic poisoning, or arsenicosis, occurs when someone takes in a high level of the substance, which is a natural, semi-metallic chemical that can be swallowed, absorbed, or inhaled. Symptoms of high levels of arsenic are stomach pain, vomiting, diarrhea, a numbness or burning sensation in the hands and feet, coma, and death. It can also cause skin darkening, lesions, and rashes. If your tap water has higher levels of arsenic, you can get a reverse-osmosis filtration system that works on a molecular level to force water through a selective membrane that weeds out the toxin. Yes, it is a toxin, and the FDA states that long-term exposure to high levels can increase your chances of skin, bladder, and lung cancers and heart disease, diabetes, and lung disease. In utero and early childhood exposure, according to the WHO, is linked to negative impacts on cognitive development and increased death rates in young adults. The WHO also states the greatest threat to public health from arsenic comes from contaminated groundwater.

Inorganic arsenic, the most toxic type, can stay in the body for a few days, and some arsenic stays for months, even longer. Arsenic can also be present in bottled water, so once again, you may be better off with your tap water and a good filter system. It can react with cells in the body and displace a cell's interior elements that literally change the way the cell functions. Arsenate, a type of arsenic, can imitate and replace the phosphate inside a cell and impair its ability to generate energy and communicate with other cells, according to Michael Paddock in the article "What Is Arsenic Poisoning?" in the January 2018 issue of *Medical News Today*. Arsenic is also present in cigarettes. A pack of cigarettes contains anywhere between 0.8 and 2.4 micrograms of arsenic.

The International Agency for Research on Cancer (IARC) has classified arsenic and arsenic compounds as carcinogenic to humans. According to the WHO, an estimated 140,000,000 people in fifty countries have drinking water with arsenic levels above the WHO guideline value. But because arsenic-induced cancer is impossible to distinguish from cancers induced by other factors, the actual magnitude of arsenic-related health dangers is probably on the lighter side of the truth.

Countries with high arsenic levels can better treat their water sources and install arsenic removal systems, but these actions are often costly and don't include the needed reductions in occupational exposure from industrial processes. At home, you can install water filters and limit the products you buy that include arsenic (yes, foods do contain it, too, so check labels), so it behooves the consumer to be proactive and not wait for government agencies to take care of the problem.

Fluoride

When it comes to water, there is an *F* word: fluoride. It has been added to our drinking water since the 1940s and approximately 70 percent of the people who are supplied water via public water systems consume this toxin, once touted to help stop cavities and fight tooth decay. The public was never asked for their consent in adding fluoridation to our water, so this is one issue that has sparked huge debate among those who support it and those who see it as a violation of trust and public consent, not to mention a persistent attempt to poison the brain and destabilize the pineal gland. But fluoride's history seemed at the time innocent enough.

According to the CDC, which considers fluoride one of the top ten greatest health discoveries ever:

A dentist in Colorado Springs, Colorado, in 1901, Dr. Frederick S. McKay noted an unusual permanent stain or "mottled enamel" (termed "Colorado brown stain" by area residents) on the teeth of many of his patients. After years of personal field investigations, McKay concluded that an agent in

Fluoride—a chemical used to prevent tooth decay—is not only in your toothpaste but also in your water supply.

the public water supply probably was responsible for mottled enamel. McKay also observed that teeth affected by this condition seemed less susceptible to dental caries [tooth decay]. Dr. F. L. Robertson, a dentist in Bauxite, Arkansas, noted the presence of mottled enamel among children after a deep well was dug in 1909 to provide a local water supply. A hypothesis that something in the water was responsible for mottled enamel led local officials to abandon the well in 1927. In 1930, H. V. Churchill, a chemist with Aluminum Company of America, an aluminum manufacturing company that had bauxite mines in the town, used a newly available method of spectrographic analysis that identified high concentrations of fluoride (13.7 parts per million [ppm]) in the water of the abandoned well. Fluoride, the ion of the element fluorine, almost universally is found in soil and water but generally in very low concentrations (less than 1.0 ppm). On hearing of the new analytic method, McKay sent water samples to Churchill from areas where mottled enamel was endemic; these samples contained high levels of fluoride (2.0–12.0 ppm).

The identification of a possible etiologic agent for mottled enamel led to the establishment in 1931 of the Dental Hygiene Unit at the National Institute of Health headed by Dr. H. Trendley Dean. Dean's primary responsibility was to investigate the association between fluoride and mottled enamel. Adopting the term "fluorosis" to replace "mottled enamel," Dean conducted extensive observational epidemiologic surveys and by 1942 had documented the prevalence of dental fluorosis for much of the United States. Dean developed the ordinally scaled Fluorosis Index to classify this condition. Very mild fluorosis was characterized by small, opaque "paper white" areas affecting less than or equal to 25% of the tooth surface; in mild fluorosis, 26%–50% of the tooth surface was affected. In moderate dental fluorosis, all enamel surfaces were involved and susceptible to frequent brown staining. Severe fluorosis was characterized by pitting of the enamel, widespread brown stains, and a "corroded" appearance.

Dean compared the prevalence of fluorosis with data collected by others on dental caries prevalence among children in 26 states (as measured by DMFT) and noted a strong inverse relation. This cross-sectional relation was confirmed in a study of 21 cities in Colorado, Illinois, Indiana, and Ohio (11). Caries among children was lower in cities with more fluoride in their community water supplies; at concentrations greater than 1.0 ppm, this association began to level off. At 1.0 ppm, the prevalence of dental fluorosis was low and mostly very mild.

The hypothesis that dental caries could be prevented by adjusting the fluoride level of community water supplies from negligible levels to 1.0–1.2 ppm was tested in a prospective field study conducted in four pairs of cities (intervention and control) starting in 1945: Grand Rapids and Muskegon, Michigan; Newburgh and Kingston, New York; Evanston and Oak Park, Illinois; and Brantford and Sarnia, Ontario, Canada. After conducting sequential cross-sectional surveys in these communities over 13 to 15 years, the incidence of caries was reduced by 50 to 70 percent among children in the communities with fluoridated water (12). The prevalence of dental fluorosis in the intervention communities was comparable with what had been observed in cities where drinking water contained natural fluoride at 1.0 ppm. Epidemiologic investigations of patterns of water consumption and caries experienced across different climates and geographic regions in the United States led in 1962 to the development of a recommended optimum range of fluoride concentration of 0.7–1.2 ppm, with the lower concentration recommended for warmer climates (where water consumption was higher) and the higher concentration for colder climates.

The effectiveness of community water fluoridation in preventing dental caries prompted rapid adoption of this public health measure in cities throughout the United States. As a result, dental caries declined precipitously during the second half of the 20th century. For example, the mean DMFT among persons aged 12 years in the United States declined 68%, from 4.0 in 1966–1970 to 1.3 in 1988–1994 (CDC, unpublished data, 1999). The American Dental Association, the American Medical Association, the World Health Organization, and other professional and scientific organizations quickly endorsed water fluoridation. Knowledge about the benefits of water fluoridation led to the development of other modalities for delivery of fluoride, such as toothpastes, gels, mouth rinses, tablets, and drops. Several countries in Europe and Latin America have added fluoride to table salt.

What the CDC history doesn't mention is Drs. H. Trendley Dean and Dr. Gerald J. Cox, both of whom began the studies in Grand Rapids, Michigan, in 1945, were employees of Andrew William Mellon, the founder of Alcoa, the Aluminum Company of America. Mellon was the U.S. treasury secretary at the time and oversaw the Public Health Service and wanted Dean to study the effects of naturally fluoridated water on teeth to prevent cavities despite the fact it caused discoloration. In 1939, Cox's own research into fluoride on the teeth of rats was enough to cause him to suggest the entire water system be fluoridated, and it was.

Banker and businessman Andrew Mellon was U.S. secretary of the treasury from 1921 to 1932. He supported fluoridation of water supplies and likely influenced Dean and Cox's studies of fluoride.

In *Population Control*, Jim Marrs writes that despite fluoride on record as causing harm to crops, livestock, and people who lived downwind from industrial plants, it was officials who had been involved in the Manhattan Project who assisted in creating a profluoride campaign to distract the public from any health concerns. The campaign involved ads and information revolving around children with shiny teeth yet never mentioned that the studies that backed the claims that fluoridation helped fight cavities were poorly designed and based on shoddy science. "By today's evidence-based medicine standards these studies do not provide reliable evidence that fluoride does indeed prevent cavities," according to Dr. Donald Miller, a retired cardiac surgeon and professor of surgery at the University of Washington, Seattle.

Yet dentists who belong to the American Dental Association (ADA) continued to insist fluoride in water, toothpaste, and mouthwashes was not only safe but also helped reduce cavities and tooth decay by 20 percent. The classic tooth discoloration from higher exposure to fluoride was passed off as a cosmetic issue and nothing to be concerned about.

A number of studies followed promoting both sides of the fluoride story, and a growing body of researchers and scientists, as well as concerned citizens, began asking if the chemical was doing more harm than good. They were labeled as wild conspiracy theorists from the get-go, but as time went on, more studies gave doubt to the original acceptance of fluoride as a wonder drug that could save the teeth of the world.

Soon, cities began to demand an end to the fluoridation of their water, citing that it was a form of forced medical practice and against moral and ethical codes. People were never asked if they wanted fluoride in their water. A study published in the May 2014 *Journal of Analytical Chemistry* by Russian and Australian chemists found that fluoride in water and toothpaste was linked to a rise in urinary stone disease. Other studies pointed to increases in kidney stones. A 2012 meta-analysis by the Harvard School of Public Health and China Medical University found evidence that fluoride negatively affected cognitive development in children, giving them much lower IQ scores than children who lived in areas with little fluoride exposure.

 A 2012 metanalysis by the Harvard School of Public Health and China Medical University found evidence that fluoride negatively affected cognitive development in children....

Approximately fifty additional studies in 2014 found a distinct relationship between lower IQ levels in children and "elevated fluoride exposure." These and dozens of other studies with humans and animals proved that exposure to higher levels of fluoride caused negative damage to the developing brains of children, including a 2006 study by the National of Academy of Sciences that resulted in recommendations to the EPA to lower admissible limits of fluoride in drinking water from 4 milligrams to 0.7. Another 2006 study by the National Research Council (NRC) determined more research was needed on the possible effects of fluoride exposure on endocrine and hormone function after finding that the toxin inhibited the thyroid iodide-concentrating mechanism. Rising rates of thyroid disease in the years following also prompted many researchers to take another look at fluoride.

A spate of newer studies, including a landmark 2017 government-funded study, found a strong link between exposure to fluoride in pregnant women and the IQ levels of their children. The higher the level of fluoride found in the urine of pregnant women, the lower the IQ of the children. A 2020 study conducted in Canada found that children who were bottle fed and who lived in communities with fluoridated water lost up to 9.3 IQ points compared to children in nonfluoridated areas. A 2019 study published in *En-*

vironmental Health linked a doubling of sleep apnea symptoms in adolescents in the United States to the fluoride levels in their drinking water, which was ascribed to the effects fluoride has on the pineal gland.

Yet another 2019 study, this one funded by the CDC and published in *Environmental International,* reported that fluoride exposure in water led to a reduction in the function of kidneys and livers in adolescents in the United States. A draft systematic review published by the National Toxicology Program in 2020 featuring 149 human studies and 339 animal studies concluded that fluoride was a presumed neurotoxin due to the large quantity, quality, and consistency of brain studies.

The antifluoride conspiracy theorists had an awful lot of science to back up their claims. They pointed to the fact that fluoride is a hazardous waste byproduct of the aluminum, phosphate, and fertilizer industries and should be labeled as a toxin and illegal to add into the water system. Many turn to unfluoridated bottled water and water filtration systems, and more cities are considering putting an end to involuntary fluoridation, but for all those who grew up with fluoride in the daily drinking water, the damage has been done.

Fluoride is also in all toothpaste unless otherwise labeled. This includes many supposedly natural brands. It is also in mouthwashes. Those who believe it is hazardous to our health point to the warnings on these products to not swallow the toothpaste or mouthwash. If it is perfectly safe, as we are told it is in our water, why a warning? Perhaps it is because it's a byproduct of aluminum and a neurotoxin, with adverse effects on the brain in high doses. This also implies that even in lower doses, long-term use can lead to the same effects.

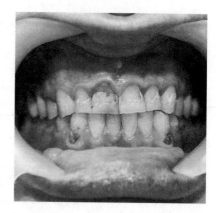

In the article "Fluoride: The Toxic Import from China Hidden in This Everyday Beverage" on Mercola.com, Dr. Bill Osmunson, a dentist with a master's degree in public health, discusses the fact that fluoride is a toxic industrial waste product and could be contaminated with lead, arsenic, radionuclides, and other contaminants.

Young children whose teeth are still forming enamel are particularly vulnerable to fluoride, which can cause damage called fluorosis.

The daily dose of fluoride recommended by the ADA is the same level that causes an eight-point drop in IQ, and even swallowing a quarter milligram of toothpaste with fluoride in it is enough to call Poison Control. A single glass of water has the same amount.

"Is it appropriate to put the substance into water, where some people may drink less than a liter a day and others drink up to 19 liters a day? That's a huge difference in the dosage amount of fluoride that they are getting," says Osmunson. He also questions whether the healthier teeth of people can really be ascribed to fluoride or to just taking better care of them. Not only that, but current levels of fluoride are two to three times the amounts first suggested way back in the 1950s.

Osmunson criticizes China for not using the toxin in its water but shipping its phosphate fertilizer to the United States, where it enters our water supply. The fluoride we have in our water is not even pharmaceutical grade but instead is a mix of contaminants that cause havoc on the body and the brain.

When it comes to IQ loss in children, Osmunson reminds us that "no one fixes IQ. This is an irreparable, irreversible damage that's happening to our public." A review of the research also indicates that fluoride might not even be the culprit in any betterment of tooth health because of flawed science and the possibility of naturally occurring fluoride working alongside high calcium and mineral levels that might also be responsible for reducing tooth decay. Some studies pointed to fluoride hardening teeth and making them able to withstand acids, but the evidence that fluoride was the cause was not strong enough, according to Osmunson.

Meanwhile, studies show that as a neurotoxin, fluoride causes increased absorption of lead, lethargy, muscle disorders, dementia, arthritis, brain damage, lower IQ, goiters, increased rates of tumors and cancers, lower thyroid function, lower fertility rates, increased attention deficit hyperactivity disorder (ADHD), cardiovascular inflammation and atherosclerosis, bone fractures, genetic damage, cell death, and a disrupted immune system. Heard enough? The negative health associations and lack of strong benefit to tooth health are enough to make you want to never drink water again. The Fluoride Action Network (FAN), based in Binghamton, New York, hopes to end water fluoridation in Canada and the United States for good, although as of today, about 60 percent of Canada's drinking water is not fluoridated. At the very least, households should be able to opt out of having any kind of toxin put into their food, water, or air without their consent. That's one of the rea-

sons those against fluoride speak out loudly and passionately, especially in light of the fact that York University in Toronto found over
ten years ago that the levels of tooth decay in children all across
Europe had decreased regardless of whether or not they had fluoridated water, with the largest decreases in countries that don't fluor-
idate at all, such as Sweden, Finland, and Denmark. Why, then, take
the chance of drinking water laced with a known neurotoxin if it
isn't doing what it has been heralded to do for over the last fifty
years or so?

 The Fluoride Action Network (FAN) ... hopes to end
water fluoridation in Canada and the United States
for good, although as of today, about 60 percent of
Canada's drinking water is not fluoridated.

In June 2020, a federal trial took place in San Francisco that
some claimed would be the beginning of the end of fluoride. FAN
sued the EPA to force it to ban or limit adding fluoride to drinking
water in the United States. FAN argued that the fluoridation violated the Toxic Substance Control Act and presented unreasonable
risks to the public, especially to children and babies in the womb.

Over 125 communities have discontinued using fluoride in
their public water systems, including Portland, Oregon, with more
cities and towns poised to do so in the coming years.

A Bigger Agenda?

Deeper down the conspiracy rabbit hole are those who ac-
cuse the government of using sodium fluoride to dumb down, sedate, and even kill off people. They reference all the scientific
studies showing neurological damage from fluoride, but they take
it a few steps further by asking if there is a more sinister agenda to
its inclusion in our water system. After all, they say, it is the primary
ingredient in rat poison and an ingredient in Prozac and the Sarin
nerve gas. They point to the fact that as little as one-tenth of an
ounce can kill a 100-pound adult and that Adolf Hitler himself used
fluoride to keep concentration camp prisoners weak and docile.

Where did all this come from, anyway? Was there a far-
reaching Nazi agenda to use fluoride as part of a mass control plan
and also a way to reduce the population of Jews and Romanis by
causing sterility in women, a known side effect of fluoride overexposure? Did the American government pick up on this practice

as a method of keeping the population oblivious to government overreach and abuses? Is the powerful Illuminati organization of the most wealthy and elite individuals on Earth behind this quest to calcify the pineal gland in the brain, also known as the "third eye," to dull and destroy the awakening of consciousness on a higher level?

The conspiracy has taken on strong legs as one of the favorites as part of talk of a depopulation agenda we will discuss in a later chapter. Might it have been the IQ studies that prompted conspiracy-minded folks to see further proof of this nefarious global plan?

There is science that backs up the claim that calcification of the pineal gland is not a good thing. This gland regulates the body's sleep-wake cycle, regulates the onset of female puberty, and protects cells from damage by free radicals. The aforementioned NRC study did find evidence that fluoride exposure caused decreased production of melatonin and other effects on the pineal function, which could trigger even more negative effects on humans. Calcified pineal glands then mean fewer properly working pineal glands; in fact, in a 1997 study, British scientist Jennifer Luke discovered that fluoride accumulates in very high levels in the pineal gland!

Luke's research also found that accumulation of fluoride reduced melatonin levels and shortened the time to puberty in animal experiments, too, prompting the NRC to state that fluoride is likely to decrease melatonin production and have other effects on normal pineal gland function in humans. The other effects on the brain mentioned in the study included neuronal degeneration, damage to the hippocampus region, reduction in nicotinic receptors, reduction in protein content, and an increase in oxidative stress. These effects were even more pronounced when in the presence of iodine deficiency and an excess of aluminum.

Nazi leader Adolf Hitler poisoned concentration camp prisoners with fluoride as a way to make them weak and easy to control.

It also might be linked to the increasingly higher number of girls in the United States who are reaching puberty age earlier, which leads to a higher risk of breast cancer. Other epidemiological studies done in 1956 and later in 1983 backed up Luke's research, stating that girls living in fluoridated communities reached puberty five months earlier.

The physiological and neurological damage to the pineal gland is backed up by science, but what about the claims of spiritual damage? The pineal gland is said to be the doorway to the Third Eye of higher levels of consciousness as well as intuition, psychic abilities, the connection to God or source, and the like. If fluoride is damaging to the pineal gland's roles in governing physical aspects of the brain and body, is it too far-fetched to say it is also damaging the doorway to higher perception, conscious awareness, and the ability to transcend the limitations of the five senses?

 The conspiracy that fluoride keeps people at a lower level of consciousness while also making them physically tired and weak may not be as wild as first thought.

The conspiracy that fluoride keeps people at a lower level of consciousness while also making them physically tired and weak may not be as wild as first thought. That's not to say Hitler had a fluoride extermination plan or that our government has picked up on that plan with full gusto, but there is no denying that fluoride is not the harmless and helpful ingredient in our water and products as once claimed. Once again, we must also revisit the question of morals and ethics and ask if it is acceptable for our government to decide, without our consent, to load up our water sources with a chemical that has been proven both toxic and carcinogenic, especially when its benefits to fighting tooth decay rest on shoddy past research and inconsistent real-world findings.

So, did Hitler and the Soviet Union's Joseph Stalin use fluoride in camps to weaken and sedate their prisoners? While there is no direct proof of this, many writers and researchers have made such allegations, including Charles Eliot Perkins, a researcher in chemistry, biochemistry, and physiology who wrote a pamphlet for the Fluoridation Educational Society in 1952 called "The Truth about Water Fluoridation." He said, "Repeated doses of infinitesimal amounts of fluoride will in time reduce an individual's power to resist domination by slowly poisoning and narcotizing a certain area of the brain and will thus make him submissive to the will of those who wish to govern him. Both the Germans and the Russians added sodium fluoride to the drinking water of prisoners of war to make them stupid and docile." Other such statements and allegations are made on conspiracy sites and in magazines and books, but there is no proof. Even if fluoride was used, we can never prove the motive behind it unless there is a smoking gun.

But at the same time, we can never disprove it.

As more cities and towns sign on to stop the fluoridation of public water, we may see an uptick in IQs even as we see a decrease in the many associated diseases and illnesses from consuming neurotoxins on a daily basis. But fluoride is just one of many contaminants in water that, though invisible to the eye, packs a powerful and visible influence on our health and well-being.

DIRTY EARTH: TOXINS IN OUR SOIL

We need healthy, nutrient-rich soil to grow our food. From the vast agricultural belts and farmlands to the grazing lands of cattle or dairy cows, the toxins in our soil compromise not only the health of the plants but also our own as we consume them. Try to grow a plant in your own garden with low-grade soil and see how healthy it ends up, if it lives at all.

Soil can be contaminated with anything from pesticides and herbicides to petroleum products, radon, lead, industrial toxins, chromated copper arsenate and creosol, farm waste, local residential waste, local auto exhaust, chemical plant runoff, and more. There is a disconnect when thinking of dirty soil because we tend to think our food is separate or that it can be washed off and made clean, but when these toxins are a part of the nourishment the plants grow from, it's hard to take the toxin out of the tomato, so to speak.

Though soil may be marked safe for walking on, playing on, or even planting our home gardens with, when it comes to our food supply, it begs a second and much closer examination.

Soil Pollution

Before we get to that, though, we need to look at the different ways humans are exposed to soil contaminants. Children ingest soil and breathe in the soil particles when playing outdoors. Contaminated soil dust also gets into the lungs of adults outside enjoying fresh air but most particularly when gardening. When we dig up soil, we disrupt volatile and biological contaminants as well as chemicals and toxins that become airborne. Windy days also kick up loose soil particles that can be absorbed through the skin or breathed in and affect the lungs' airways.

Contaminated soil gets into our food and our air. Whenever we dig into the soil, the dust is carried by the wind and into our lungs.

Soil can include asbestos fibers, lead, and a host of other chemicals that harm human health when ingested. Creosote is absorbed into the skin and can cause rashes and blistering.

Covering stretches of bare soil with plants can help reduce the toxin levels in the soil, but of course the plants themselves absorb the toxins, so it's best to not plant foodstuffs in dirty soil. Instead, you can use outdoor plants that won't be eaten by humans, butterflies, or bees to protect their health.

Obviously, the worst type of soil contamination is that which enters our bodies through the food and crops we grow. Sometimes, the toxins are a part of an existing site, such as the site of a former industrial complex, an old building, or a manufacturing plant that has been turned into a community garden. The soil can be treated, but it's expensive. But if that is the only soil available to grow food in, it's a necessity. Many vegetables and herbs absorb some of the contaminants to get them out of the soil but then present a greater danger when consumed. Garden beds can be lined with chemically treated wood or other products, so those can cause the leaching of chemicals into the soil.

It's getting harder and harder to find pure, healthy, nutrient-rich soil to grow our food supply with because of overuse and pesticides, as well as contaminated air and water that contribute via rainfall, water runoff, or irrigation methods. It's hard to escape the contaminants in our soil because even if you find good land to grow on, you must deal with wind blowing contaminants onto your soil. Soil remediation is expensive, and for smaller gardens, many people turn to bagged soil that claims to be toxin free and heavy on the nutrients.

Any gardens or crops grown near construction sites must deal with the potential of petroleum products and paint, cleaners, solvents, and other toxic agents used to build homes and offices. Fibers from roofing products are carried in the wind and drop to the soil. Landscaping introduces pesticides and herbicides that are heavily linked to health problems in humans. Even fertilizer, used to help crops and plants grow, can become a toxic agent when it enters the human body.

 Even fertilizer, used to help crops and plants grow, can become a toxic agent when it enters the human body.

Soil pollution is a lesser concern to most people than air or water because of the huge disconnect between the food we buy at the local grocery store and the process by which it was planted, grown, and then distributed. We see only the result and don't worry about industrial contamination, agricultural waste and runoff, acid rain, oil and chemical spills and leakages, garbage and waste disposal, and the use of herbicides and pesticides because we think in the course of that food reaching our homes, it was probably cleaned and protected. We then go home and wash it again if produce, and it magically is clean and toxin free. Meat products with animals raised on toxins in their feed don't even faze us if the label reads "grade A" or "organic."

Rotten soil makes rotten food makes rotten feed. Decreased fertility of the soil leads to decreased plant and crop yields, which leads to decreased nutrients, which leads to more pesticides to protect what does grow or the use of genetically modified organisms (GMOs), which we will discuss later. All of these sources of pollution change the structure of the soil and sadly the sources are mainly man-made, which means we can control, regulate, or end them, but because of the powerful influences of industry lobbyists on our government officials and agencies, they fail more than they succeed.

In 2020, a documentary was released titled "The Need to Grow" made by the Food Revolution Network, stating that our planet's farmable soil was running out at an alarming rate and would be gone in sixty years if no actions were taken to prevent such a catastrophe.

Pesticides

When it comes to soil contaminants, pesticides are the biggest threat to human health. Pesticides are sprayed on crops and plants and then absorbed into the soil but not before covering fruits and vegetables with a layer of toxicity that has been linked to cancer and other diseases. The use of pesticides has a beneficial purpose—to rid plants of bugs, rats, mice, critters, and anything that can keep the produce we eat from growing fully and abundantly. Without pesticides, we lose a lot of produce and food to insects and other pests, causing less available food and a loss in profits for farms and agricultural industries.

Even in your home garden, you probably spray with a pesticide from a home-improvement store unless you're savvy enough to know how to make your own that is not toxic to humans or animals. But chances are, like farms and agricultural companies, you go to the store and buy a product already in existence without asking what the ingredients are and how they may affect your body and health. Pesticides are sprayed on food in grocery stores and in food storage facilities and in some cases sprayed throughout neighborhoods via planes to stop mosquitoes that carry malaria, yellow fever, and the West Nile virus.

Does this photo make you want to eat the vegetables in this crop? If pesticides are so dangerous that you need protective gear to use them, how safe do you think it is to have such chemicals leaching into the soil or absorbed into your food?

Pesticides are described as chemical or biological agents designed to kill, deter, or discourage pests from something, such as a plant. There are a number of plant-eating creatures, such as birds, worms, insects, pathogens, fish, humans, and microbes, and they not only destroy the organism but can also spread disease as vectors or carriers. Pesticides are meant to get rid of these pests and deter and destroy the growth of weeds and undesirable plants. There are different kinds of pesticides, such as herbicides, insecticides, fungicides, animal repellents, insect repellents, piscicides (fish), avicides (birds), ovicides (kill eggs of pests), rodenticides, slimicides (kill slime-producing algae, molds, etc.), bactericides, and molluscicides. These contain chemicals that focus on specific pests and target them with key ingredients that may not work against other types of pests.

A weed killer may not work on getting rid of rats, and rat poisons won't stop birds from eating a plant or crop. Pesticides are often dispersed in a hydrocarbon-based solvent-surfactant system and usually work by poisoning the pest in question. Newer pesticides take advantage of the mechanisms that allow a plant's own natural defense chemicals to be released and could be safer than the older chemical formulas. The first known use of a pesticide traces back to elemental sulfur dusting in ancient Sumer around 4,500 years ago in the Mesopotamian region. By the fifteenth century, arsenic was the favored chemical as well as lead and mercury for keeping pests from destroying crops.

 The first known use of a pesticide traces back to elemental sulfur dusting in ancient Sumer around 4,500 years ago in the Mesopotamian region.

In the seventeenth century, nicotine sulfate was extracted from tobacco leaves, and the nineteenth century introduced natural pyrethrum, derived from the chrysanthemum flower, and rotenone, from the roots of tropical vegetables. Up until the 1950s, arsenic was the main ingredient in pesticide use, but when DDT was discovered by Paul Miller, it took over as the dominant form until the mid-1970s when organophosphates and carbamates took over.

Through the 1960s, herbicides were popular, including one that would become the subject of incredible debate over its toxicity and use in food: glyphosate. As pesticides were used more frequently, pesticide resistance began to rear its head. In the 1940s, only about 7% of crops were lost to pests. By the 1980s, that percentage had risen to over 13% and today, there are up to 1,000 species of insects and weeds alone that have developed pesticide resistance.

A number of pesticides are grouped together into chemical families such as organochlorines, carbamates, or organophosphates. One of the most notorious pesticides of the past, DDT, falls into the category of organochlorine hydrocarbons.

DDT

There was a time when dichlorodiphenyltrichloroethane (DDT), an organochlorine that fights malaria, was sprayed in houses on the walls and on children without face protection. DDT is a synthetic organic compound and a chlorinated aromatic hydrocarbon that was used widely as a powerful insecticide until it was banned in several countries, including the United States by the EPA as one of its first official acts. DDT was linked to damage to the environment and to human and animal health until 1972.

First synthesized by Austrian chemist Othmar Zeidler in 1874, it wasn't until 1939 that Swiss chemist Paul Hermann Müller discovered DDT's insecticidal properties. While working for the J. R. Geigy firm, he was able to demonstrate that DDT killed the Colorado potato beetle, which was at the time destroying potato crops in the United States and Europe. During World War II, DDT was used to control body lice, plague, malaria, and typhus among troops and civilians. In May 1963, author and environmentalist Rachel Carson, who wrote the seminal book *Silent Spring*, appeared before the U.S. Department of Commerce asking for a pesticide commission and for regulation of DDT. Ten years later, this commission changed its name to the Environmental Protection Agency (EPA), and it immediately banned DDT. Carson became the hero of the modern environmental movement as a result.

Biologist and conservationist Rachel Carson's 1962 book, Silent Spring *helped kick off the international environmental movement.*

Sadly, in the 1940s, children had been exposed to high levels of the toxin when sprayed to kill off mosquitoes. According to *New Scientist* magazine's August 2018 article by Andy Coghlan, "Exposure to Insecticide DDT Linked to Having a Child with Autism," a Finland study showed mothers with high DDT exposure in their blood were more likely to have autistic children. Children in the United States were also sprayed with DDT, which many thought was a miracle cure for polio. Public swimming pools, city streets, and entire villages were openly sprayed without concern for what the toxin might do to human health.

Later studies showed that side effects from DDT exposure led to vomiting, tremors, dizziness, and seizures and is now thought to be a human carcinogen. The insecticide is still used today in countries in Africa, Asia, and South America. It is still legal to manufacture DDT in the United States, which exports it to countries that can legally use it, but it's also made in China, India, and North Korea.

 In 1996, South Africa stopped the use of DDT, and the malaria cases skyrocketed in one province alone from 8,000 to 40,000....

DDT lasted so long because it was fairly cheap to make, and it worked effectively. Some people argue that banning DDT killed more people than it saved because it allowed them to die of malaria, yellow fever, and other diseases. The National Academy of Sciences in 1970 stated that the use of DDT may have saved 500,000,000 lives, but at that time, there was no knowledge of the untold future damage it would do to those it was sprayed on, including children and babies. Pro-DDT activists continue to point to the widespread life-saving effects of the toxin and point to the reductions of malaria deaths across the globe, which, according to the World Health Organization (WHO), infects over 300,000,000 people per year, most of them African children. They point to the banning of DDT in Sri Lanka where in 1948, there were 2,800,000 cases and 7,300 deaths. With DDT use, the malaria cases fell to 17 and zero deaths in 1963. When DDT was discontinued, the infection rate rose back up to 2,500,000. In 1996, South Africa stopped the use of DDT, and the malaria cases skyrocketed in one province alone from 8,000 to 40,000, and by the year 2000, there was a 400 percent increase in deaths. DDT is now back in use, and the region has seen a drastic reduction in deaths.

So, do you continue to use a known toxin that damages the environment and human beings, as some studies have shown it to also be linked to lymphoma, pancreatic cancer, breast cancer, and liver cancer, or do you continue to use a chemical that saves people from other deadly diseases such as malaria? According to the Pesticide Action Network, DDT breakdown products were found in the blood of 99 percent of people tested by the CDC, and those same breakdown products were found in 42 percent of all greens, 60 percent of all heavy cream samples, and 28 percent of all carrots tested by the U.S. Department of Agriculture (USDA).

In areas with mosquito-borne malaria, the disease not only kills people but negatively affects the economy. It's an intriguing and morally explosive argument, but there are safer alternatives that these countries can use as their insecticides of choice. The problem is often cost and speed of efficiency, and for poorer nations, cheaper means better when trying to do what they believe is the right thing for their people.

In the United States, DDT spraying is a thing of the past, although the repercussions to the health of those exposed is a present concern. A story broke in November 2020 about a secret DDT dumpsite identified by deep-sea robotics, revealing potentially 500,000 barrels of DDT waste off the coast of Los Angeles, California. Photos taken by these robots correspond to reports of thousands of such barrels dumped into the ocean in the months following World War II and that the entire area might contain between 336,000 to 504,000 of acid sludge waste contaminated with DDT residues. Why off the coast of California? Because the state was the home of Montrose Chemical Corp., which opened a plant to manufacture DDT in 1947 near the city of Torrance. This site is considered one of the most hazardous in the nation, even today.

These barrels may be sending out DDT-contaminated waste into the ocean, and this waste has shown up in California bottlenose dolphins. If this waste is down there oozing out into the open waters, it has no doubt already entered the bloodstreams of living things in the ocean … and, eventually, us.

Chlorpyrifos

In 2015, the EPA under President Barack Obama proposed a federal ban on a pesticide called chlorpyrifos. Three months after President Donald Trump was sworn in, that ban was lifted, although Hawaii and California both banned the toxin since. This despite numerous studies showing the toxin was not only damaging to human health but was also washing into streams and rivers and killing off wildlife, including endangered salmon.

Chlorpyrifos is a deadly neuro-toxic pesticide used for over fifty years in the United States to protect corn, wheat, apples, and citrus crops. Dow Chemical Company introduced the pesticide for use in homes and fields in 1965, and it was highly effective in deterring pests of all kinds. It was sprayed not only around homes and crops but also in and around offices, on golf courses, and throughout municipalities, including public parks where children played.

Estimated Agricultural Use for Chlorpyrifos 2011
EPest-Low

Estimated use on agricultural land, in pounds per square mile

In 2011 the use of chlorpyrifos, a pesticide neurotoxin, was still widespread in the United States. It was banned in 2015 but brought back under President Trump's watch.

It is also known as a nerve agent with such side effects as brain damage to children, memory loss, attention deficit disorder (ADD), and reduced IQ. It is especially damaging to farmworkers who are exposed to higher levels when the fields they work in are sprayed, even for days afterward, and to children, but it has managed to find its way into the water and food supplies all over the country. It is sprayed on crops to kill a variety of pests and is harmful when touched, eaten, or inhaled. It has the smell of rotten eggs and causes acute poisoning by suppressing the enzyme that regulates nerve impulses. Other side effects of exposure include seizures, delayed motor development, lower birth weights, convulsions, respiratory paralysis, autism, and death.

Children suffer higher chlorpyrifos exposure because they drink more water and juices and consume more fruits and veggies tainted with the pesticide, and they also tend to touch their eyes and mouths more than adults. In the fall of 2020, Corteva (formerly Dow) was hit with two lawsuits over the pesticide linked to brain damage in children. The parents claimed the pesticide caused their children's injuries and that the company knew the product could cause brain damage in children, including unborn children whose mothers came in contact with the chemical. The suits were both filed in California Superior Court one year after Corteva stated it would no longer use chlorpyrifos after it was banned in the state.

The first case involved a child who suffered severe neurological injuries as a result of in utero, infant, and ongoing exposure to chlorpyrifos. The second child was the daughter of farmworkers and was diagnosed with autism, obesity, and vision problems by the age of thirteen.

In November 2016, the EPA released a statement revising the health risks from this pesticide by confirming there were no safe uses for it. The EPA suggested a ban, but to this day, the pesticide is not banned in the United States except in Hawaii and California, thanks to stonewalling and heavy lobbying from the manufacturer, Dow Chemical, beginning in 2016. According to Children's Health Defense, the EPA is one of many federal agencies that are dominated by industry lobbyists and contributors that keep any safe regulations from passing into law, calling them a "hall of funhouse mirrors" where organizational goals get distorted and twisted from their original form until you wonder what was real in the first place. Among other agencies that continuously cater to corporate interest over human health are the FDA, USDA, CDC, and the National Academy of Sciences, making it incredibly difficult to get changes made that benefit people over profits.

The Children's Health Defense also pointed to the acceptance and reliance of the EPA on clearly biased studies that downplayed the toxicity of this and other pesticides rather than the multiple independent studies that show harmful neurotoxic effects, especially on developing nervous systems. Instead, the industry-backed and funded studies were given more attention during the regulatory review process. The Children's Health Defense stated that "flagging exposure to chlorpyrifos and other organophosphates as a risk factor for autism spectrum disorders (ASD) and other developmental problems … concerned scientists [who] have been sounding the alarm for quite some time. Yet, despite the substantial body of evidence documenting adverse effects not just on human health but also on wildlife and the environment, the Environmental Protection Agency proclaimed … that it will take no action other than to 'continue to review the safety of chlorpyrifos.'"

Washing your fruit is not, unfortunately, enough to get rid of the residues of some pesticides that are sprayed on produce.

Fortunately, in April 2021, the Ninth Circuit Court of Appeals ruled that if the EPA could not define what a safe level was for children, it would have to ban chlorpyrifos from use on food crops.

Of all the pesticides on the market, one in particular angers more people than any other for its pervasive presence in our food and water.

It's the Law ... No?

After the release of a 1993 report by the National Academy of Sciences, Congress made stronger protections for children from toxic pesticides, even criticizing the EPA for failing to address how children were especially and uniquely susceptible to the toxic damages of exposure.

Congress then unanimously passed the Food Quality Protection Act in 1996, which required the EPA to protect children from harmful pesticides and to ensure that these products won't harm infants or children from aggregate exposure, no matter the cost to industry.

Sadly, the ban on chlorpyrifos exposed the deep flaws in enforcing this act when it was overturned by the Trump administration despite the clear and present dangers chlorpyrifos presented in many scientific studies to children, adult farmworkers, and wildlife. It was banned from household use in 2000 but continues to be sprayed on crops and in agricultural areas.

So much for the law.

Glyphosate

Glyphosate is an herbicide that is applied to the leaves of plants to kill broadleaf plants and grasses. In its sodium salt form, it is used to regulate plant growth and ripen crops. It is the most widely used herbicide in the United States since its first registered use in 1974 and is used both residentially and industrially to kill weeds, protect lawns and gardens, and assist agricultural and forest growth. It is even used to control specific aquatic plants.

It comes in the form of acid as well as salt, and it can be solid or liquid. Over 750 products contain glyphosate in the United States alone. It is called a nonselective herbicide because it kills most plants exposed to it by preventing the plant from making necessary growth proteins. It does this by stopping an enzyme pathway called the shikimic acid pathway that is needed by plants and many microorganisms to grow properly.

With more widespread use, it became clear that exposure to glyphosate on the skin or to the eyes and airways, or by not washing hands after touching a plant sprayed with the toxin, can cause great harm to human health. Because it does not vaporize once sprayed, it remains as a liquid on plants and makes exposure to humans and animals much easier.

Industry Claims

If you ask the pesticide industry about the toxicity of glyphosate, you will be told it doesn't easily pass through human skin, if absorbed or ingested it will pass through the body quickly, and that it leaves the body via feces or urine without being changed into another chemical. You will also read that animal and human studies by regulatory agencies in the United States, Canada, Japan, the United Kingdom, and Australia show it is not likely to be carcinogenic after being fed in high doses to laboratory animals, though they will admit to the possibility of it being carcinogenic based on the findings of a committee of scientists working for the WHO's International Agency for Research on Cancer (IARC).

The industry also claims children are not harmed by exposure to glyphosate, as per the Food Quality Protection Act, that it is not likely to get into groundwater because it binds lightly to the soil, and that pure glyphosate is low in toxicity to fish and wildlife, instead suggesting toxicity may come from other ingredients in products containing glyphosate.

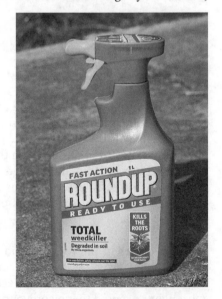

Rounding Up Roundup

In 2018, the lawsuits against Monsanto's weed-killer Roundup began. A terminally ill California man won a milestone case against the agricultural giant, which merged recently with Bayer, after he claimed that Roundup gave him cancer. Dewayne Johnson, forty-six, was awarded $289,000,000 in damages after a jury ruled the company's product, which

The popular weed killer Roundup uses glyphosate, an herbicide that many claim is a carcinogen for humans and animals.

contained glyphosate, was the reason he developed non-Hodgkin's lymphoma, according to the Associated Press. Johnson used Roundup regularly when he was a school groundskeeper in the San Francisco Bay Area school district (which means the children at those schools were also exposed to cancer-causing herbicides and pesticides).

Of course, Monsanto denied the claims that its product causes cancer and appealed, resulting in a lessening of the awarded money. But the floodgates were open, and Monsanto was faced with more lawsuits. Edwin Hardeman was awarded $80,000,000 in a lawsuit linking Roundup with his diagnosis in 2015 of stage 3 non-Hodgkin's lymphoma. In 2019, Jason Laurianti sued Monsanto over his own non-Hodgkin's lymphoma diagnosis after claiming to use Roundup for over fifteen years. Again, both Monsanto and the EPA claimed the product was safe to use as directed.

There are, to date, over 11,000 additional lawsuits pending against the product and its powerful maker, Monsanto (now Bayer/Monsanto).

Glyphosate Danger

The active ingredient in Roundup, and other herbicides under fire for carcinogens, is glyphosate, which has been the farming go-to herbicide for four decades, without much attention until the IARC concluded it was "probably carcinogenic." After this, many studies were conducted mirroring that statement, some even stating outright that this particular ingredient went beyond just causing cancer. It was downright deadly.

While the EPA, the European Food Safety Authority, and other pro–big agriculture organizations insisted it was unlikely to increase cancer and, in the case of Roundup, defended the herbicide by stating glyphosate was only one of several potentially dangerous ingredients, the studies to the contrary piled up (along with the lawsuits). In "How Toxic Is the World's Most Popular Herbicide Roundup?", even Monsanto's own chemistry safety and outreach lead, William Reeves, told *The Scientist* in its February 6, 2018, issue that he wasn't surprised at studies pointing to glyphosate as a potential risk factor. He stated that the surfactants in Roundup are similar to other household products that cause "membrane degradation and subsequent mitochondrial breakdown in high doses." He equated it to the same effects as using high doses of dish detergent. Because more pesticides and herbicides have been used over the years, both improper use and overuse are definitely huge concerns for any chemical product.

Chronic exposure to a chemical through skin, breathing into lungs, and consuming in food, especially once that chemical has accumulated in the environment, is of special concern to researchers. Another concern, pointed out in the article above, is the lack of transparency of ingredients used. "A problem for scientists investigating the physiological activities of pesticides is that herbicide-producing giants including Monsanto, Roundup's developer, or Syngenta, which produces the glyphosate-containing herbicide Touchdown, aren't required to make their ingredients lists public."

Chronic exposure to a chemical through skin, breathing into lungs, and consuming in food, especially once that chemical has accumulated in the environment, is of special concern to researchers.

In the United States and Europe, companies are required to print the active ingredient the product contains on the package itself, but this is not the case for other ingredients considered inert because they don't contribute to the "herbicidal activity" of the product. Vanessa Fitsanakis, a neurotoxicologist at Robson Forensic, says "that makes it very difficult for a toxicologist to test the different ingredients to figure out what's the most toxic, or what's contributing to it." Does this let glyphosate off the hook?

According to a meta-analysis reported on in the February 14, 2019, issue of *Earth Island Journal*, this comprehensive review of epidemiological studies published between 2001 and 2018 tracked the health of nearly 65,000 agricultural workers and found a compelling link between exposure to glyphosate and increased risk for non-Hodgkin lymphoma. The findings of this review corresponded with the earlier classification of glyphosate as a potential carcinogen by the IARC. Bayer, the new owner of Monsanto, claimed the study was flawed and that it cherry-picked data, and it was noted this new study contradicted earlier assessments from the EPA to the contrary, but senior researcher Lianne Sheppard, a professor in the Environmental and Occupational Health Sciences and Biostatistics departments at the University of Washington, stated, "Our analysis focused on providing the best possible answer to the question of whether or not glyphosate is carcinogenic. As a result of this research, I am more convinced that it is."

A February 2020 analysis of 59 pest-assessment risk studies was published in the journal *Environmental Health*, which found that the EPA did not apply the 1996 Food Quality Protection Act safety factor to more than 85 percent of nonorganic phosphate pesticides. The EPA claimed it makes decisions about applying this safety factor based on a variety of peer-reviewed research, and if it decides not to, it is basically saying that the pesticide is not harmful to infants and children differently from adults. But *Consumer Reports* research scientists have accused the EPA of not looking at the overall spectrum of harm and that it doesn't use the latest science and scientific techniques to look for potential pesticide harms. In the October 2020 issue of *Consumer Reports*, the article "Stop Eating Pesticides" by Catherine Roberts states that the CR team of safety experts believe glyphosate should be banned, along with weed-destroying dicamba. The article also emphasizes the compound dangers of using multiple pesticides together and laments that only the active ingredient is studied and/or regulated.

 The findings of this review corresponded with the earlier classification of glyphosate as a potential carcinogen by the IARC.

Brenda Eskenazi, Ph.D., the director of the Center for Environmental Research and Children's Health at the University of California at Berkeley, states that most studies evaluate only the effects of one type or class of pesticide. "What we should be looking at is the whole swimming pool of chemicals that we're exposed to."

In August 2020, Mexico announced that it would be 100% glyphosate-free by 2024. The country's Environment Ministry declared that it would break the grip of Monsanto's hold by gradually reducing the amount of glyphosate farmers could use, completely phasing out the chemical in 2024. Hopefully, other nations will follow in Mexico's footsteps.

Other Dangers

The same large analysis found glyphosate might also potentially act as an endocrine disrupter after it was found in a long-term French animal study by Gilles-Éric Séralini, Michael Antoniou, and their colleagues to alter sex hormone production in male and female rats. The Mayo Clinic reports that after four decades of pervasive use of glyphosate-based pesticides, over 100,000,000 Americans now have liver disease, some as young as eight years of age.

Glyphosate is also linked to kidney damage. A January 2020 study published in the journal *Environmental Pollution* identified glyphosate in the urine of 11.1 percent of infants and young children tested, and this rose to 30 percent among newborns. Though no direct link was found with these markers and kidney damage, the study was small and limited, and the researchers suggested further studies with larger sample sizes. The study also suggested the findings in children did not apply necessarily to adults who have been exposed over longer periods of time.

Documented in the article "Glyphosate Is a Primary Cause of Kidney Damage" in the January 28, 2020, analysis by Dr. Joseph Mercola, the researchers stated, "Previous studies have associated glyphosate exposure with changes in renal function, kidney injury, and chronic kidney disease of unknown etiology. There is growing evidence linking glyphosate exposure with the epidemic of chronic kidney disease of unknown origin in farmworkers in Central America, Sri Lanka, and Central India."

In 2014, Dr. Sarath Gunatilake, professor of health science at the California State University at Long Beach, and Channa Jayasumana, Ph.D., professor of medicine and allied sciences at the Rajarata University of Sri Lanka, published papers making this link and hypothesized that consumption of water contaminated with glyphosate could have contributed to chronic kidney disease by facilitating the transport of heavy metals such as arsenic and cadmium into the kidneys. The following year, they looked at heavy metal concentrations and glyphosate in well water and found that those who drank from wells had a fivefold increase in their risk of chronic kidney disease with unknown etiology (CKDu). Their research was considered significant enough to earn them recognition from the American Association for the Advancement of Science, the world's largest scientific society and publisher of such journals as *Science*.

GMO crops are designed to be pest and disease resistant, but what does genetic manipulation do to the nutritional value of a food?

Roundup and GMOs

There is the assumption that Roundup is only sprayed on GMO crops, which are also called Roundup-ready crops because they have been designed to withstand glyphosate. But this is not true. Glyphosate is used on non-GMO crops such as wheat, oats, edible beans, corn, soy, and other crops usually right before they are harvested. The genetically engineered crops are supposed to be able to withstand the harshness of the herbicide, but what about regular crops that don't have that built-in shield?

Farmers spray the herbicide at the end of a growing season to allow for an earlier harvest and do it more efficiently and at less cost, according to the Glyphosate Renewal Group. Even organic crops are tainted with the herbicide. It's used on more than seventy of the most popular crops, so it's hard to avoid it. Even if organic farmers don't spray it directly, glyphosate is in the water supply they use to irrigate their crops. The chemical particles can settle on dust and soil miles away from their origin points. Eating organic can certainly lower the amount of exposure, but it doesn't stop exposure.

⚠ It's becoming harder to escape the presence of toxins in our food, water, and air, and with the growing proliferation of GMO products, we probably never will.

Crops sprayed with glyphosate include apples, almonds, grapes, oranges, figs, watermelons, celery, cherries, kiwis, lemons, lettuce, olives, onions, cotton, cucumbers, squash, plums, and pumpkins. In fact, it might be hard to find one crop *not* sprayed with the stuff. Nonorganic crops are more likely to be doused with pesticides of any kind, so buying organic is always best or shopping from local farms and farmer's markets where you can ask the farmers exactly what they use on their crops.

Even consuming meat and dairy means having to watch out for glyphosate as they often are fed a diet based on crops like corn that are sprayed with the chemical. It's becoming harder to escape the presence of toxins in our food, water, and air, and with the growing proliferation of GMO products, we probably never will.

With only a handful of massive agribusiness corporations in control of pesticide use, GMOs, and food supply, the industrialization of our agriculture has increased the burden on the ecosystems and overall environment. Agrochemicals such as pesticides and herbicides are rampant—in the air, water, soil, and food. As we learn more about their negative and even deadly effects on humans and animals, no doubt there will be greater pushback from these companies because their main concern is profit. It will be up to citizens and activists to push for more transparency on labels and fewer toxins used in food production. The need for a reduction in these toxins is urgent as we approach a point of no return with farmable land, drinkable water, and breathable air, and there is only so much damage our bodies can take.

GMOS AND THE MONSANTO MONSTER MACHINE

Just as the media is run by a handful of big corporations, so is agribusiness. A handful of giant pesticide and chemical companies dominate the agricultural input market not just in the United States but on a global scale as well. These companies own the world's seeds and run the world's genetically modified organism (GMO), pesticide, and biotechnology industries. According to the United Nations Conference on Trade and Development, this kind of corporate concentration of the market has dangerous and far-reaching consequences for consumers and food security as they continue to privatize and patent every possible agricultural innovation from GMOs to new tech and pull the rug out from under smaller farms and traditional seed-growing companies. The corporations are:

▲ BASF

▲ DuPont

▲ Dow Chemical Company

▲ Syngenta

▲ Bayer/Monsanto

All of them pose the threat of monopolization as mergers occur, and the biggest merger of all so far was that of Bayer and Monsanto.

According to SourceWatch, Monsanto is the most controversial of these companies; in 2018, it merged with Bayer, the German pharmaceutical and chemical giant. Monsanto produces genomics and herbicides for a variety of crops, such as corn, cotton, oil seeds, and vegetables, and a number of branded lines of seeds, but it also has the negative distinction of having produced Agent Orange, PCBs, DDT, Aspartame, and the recombinant bovine growth hormone (rBGH). It is considered the biggest of the "global bullies" in the Big-Agra schoolyard.

John F. Queeny, a purchasing agent for a drug company, started Monsanto Chemical Works in 1901. (*Monsanto* was Queeny's wife's maiden name.) The goal was to manufacture saccharin, the synthetic sweetener that was only made in Germany at the time. By 1902, Monsanto was up to full-scale production, adding caffeine and vanillin to its product line. By 1915, it had sales worth $1,000,000, thanks to having the Coca-Cola Company as a major customer, and by 1917, it began producing aspirin. In 1928, the company was passed on to Queeny's son, Edgar M. Queeny, who took things to new levels, turning Monsanto into an industrial giant before he retired in 1960.

Monsanto was producing styrene products, which were vital to the World War II war effort, and purchased pharma company G. D. Searle and Co., makers of NutraSweet®, in 1985. The company sold off its sweetener business in 2000.

 Monsanto's involvement with genetically modified seeds and organisms gave it a black mark that never went away, as evidence revealed dangerous chemicals in its products, including Roundup.

During the 1990s, the company acquired several firms, such as Calgene, Inc. and DeKalb Genetics Corporation, catapulting its leadership status in the biotech world, mainly in the arena of genetically modified crop seeds. In 1994, it began the production of bovine somatotropin (bST), a synthetic supplement for dairy cows, merged with Pharmacia & Upjohn in 2000, and went public in 2002, after which it acquired various seed companies and software firms.

Monsanto's involvement with genetically modified seeds and organisms gave it a black mark that never went away, as evidence revealed dangerous chemicals in its products, including Roundup. The German pharma giant Bayer purchased Monsanto in 2018 for

Monsanto's Indestructible Disney Attraction

Monsanto's House of the Future was an attraction at the Disneyland amusement park in Anaheim, California, from 1957 to 1967.

Many people had their first introduction to the Monsanto name via rides in Disneyland's Tomorrowland, which featured the virtues of plastics and chemicals, including the House of the Future, which was made of plastic and claimed to be biodegradable. It wasn't. From 1957 to 1967, over 20,000,000 people visited the plastic house until it was torn down by Disney, or, at least, Disney attempted to tear it down. Wrecking balls bounced off the glass-fiber materials, and torches, jackhammers, and chainsaws failed to take the house down. Finally, Disney had to resort to choker cables to break off parts of the house, which were then hauled away in bits. The only thing that survived was the synthetic Astroturf used in the yard.

approximately $63 billion, and shortly after that, the flood of lawsuits began against Roundup, costing Bayer over $10 billion in settlement payouts. Monsanto officially became a part of Bayer's crop science division and no longer existed by its former name, but its long history of bullying farmers and domination of the agriculture industry remains, which culminated in a massive global protest in 2013 called the March against Monsanto.

Though all of the Big-Agra and pesticide companies had their share of disasters, deadly products, and lawsuits, Monsanto lived up to its reputation as the evilest corporation in the world. Here is a timeline of its misdeeds:

▲ 1901–1910: Saccharin, the company's primary product, was considered a poison by the government, which sued to stop production. Black mark #1. The government failed to stop the global poisoning of products, such as soda, containing saccharin.

▲ 1920s: Polychlorinated biphenyls (PCBs), once considered a miracle chemical, are manufactured and later recognized as one of the most toxic chemicals in existence. They are banned after fifty years, but not before poisoning people and the environment and causing a number of fetal deaths and immature births in East St. Louis, Illinois, where its manufacturing plant was located.

▲ 1930s: Monsanto begins production of a slew of toxic products, including industrial cleaners, synthetic rubbers, plastics, and detergents. It also creates the first hybrid seed for corn.

▲ 1940s: Monsanto is involved in the creation of the first nuclear bomb for the Manhattan Project during World War II at its Dayton, Ohio, facilities. Its research into uranium is used to later bomb Hiroshima and Nagasaki, Japan, killing hundreds of thousands of Japanese civilians as well as Korean and U.S. military servicemen. The company expands production of its highly toxic pesticides.

▲ 1960s: Monsanto teams with Dow Chemical to create Agent Orange for use in the Vietnam War, contaminating over 3,000,000 people, killing 500,000 Vietnamese civilians, causing 500,000 babies to be born with severe birth defects, and poisoning and killing thousands of U.S. servicemen who continue to suffer today. Monsanto enters into a joint venture in 1967 with IG

Farben, a German chemical firm that served as the financial center of the Nazi regime and supplied Zyklon B to the extermination camps during the Holocaust.

▲ 1970s: Monsanto's aspartame product is revealed by the Food and Drug Administration (FDA) to cause tumors and holes in the brains of rats before killing the animals. Monsanto partner G. D. Searle claims the product is safe. The investigation is dropped when Donald Rumsfeld, who would later serve as secretary of defense during both the Gerald Ford and George W. Bush administrations, becomes CEO of Searle, and Samuel Skinner, who would later serve as secretary of transportation and chief of staff during the George H. W. Bush administration, resigns from the U.S. Attorney's office to work for Searle's law firm. But the controversy rages over aspartame for another fifteen years or so before it is banned. Monsanto plants in Georgia, Idaho, Illinois, Alabama, and Virginia are labeled Superfund cleanup sites by the Environmental Protection Agency (EPA) for being so contaminated and full of hazardous waste they require special priority cleanup.

▲ 1986–2007: Numerous legal actions against Monsanto for workplace and environmental safety violations are conducted, resulting in the company paying millions in damages and settlements.

▲ 1990s: Monsanto faces a number of state and federal laws that attempt to force it to stop dumping dioxins and pesticides into drinking water systems. The company is sued by plant workers and citizens living near dump areas, resulting in $100,000,000 being paid out, a small price to pay for its huge profits gained from toxic products.

▲ 1994: The FDA approves the rBGH. The next year, Monsanto begins producing GMO crops tolerant to its notorious herbicide, Roundup.

Monsanto is one of the most powerful chemical corporations in the world and is responsible for such dangerous products as aspartame, Agent Orange, and Roundup.

Monsanto Poisons an Entire Town

The residents of Anniston, Alabama, were poisoned by PCBs produced by the Swann Chemical Company factory there that was later run by Monsanto.

Anniston, Alabama, has a lot of sick residents. Over the years, Monsanto Industrial Chemicals admitted to poisoning residents with PCBs, according to "Monsanto Poisoned This Alabama Town—And People Are Still Sick" by Harriet Washington for the July 26, 2019, edition of BuzzFeed News. The article is an excerpt from her book *A Terrible Thing to Waste: Environmental Racism and Its Assault on the American Mind* and documents the decades of toxic exposure this city of 24,000, which is 52 percent African American, endured at the hands of the agribusiness giant.

In 1935, Monsanto Industrial Chemicals took over Swann Chemical Company, which produced PCBs, in Anniston. By 1969, the plant was the city's major employer and was "discharging 250 pounds of PCBs into Snow Creek, at the heart of the city's black residential community, every day." Shortly afterward, the townspeople were battling a "legion of ailments from cancers to memory loss, confusion, and a slew of other intellectual problems. Children's behavioral problems snowballed in Anniston, along with rates of attention-deficit disorder and poor school performance." The media focused outrage on the more sensationalistic cancer clusters and lung and mobility ailments caused by PCBs, but the behavior ailments were just as pervasive, wreaking harm on the brains of infants and children.

Many people died, and many more suffered the results of the PCB poisoning and continue to do so. They played and waded in water, unaware that it was toxic, and had no idea the soil they grew plants and food on was poisonous. In fact, the article states that the residents of Anniston didn't really learn just how poisonous their city was until a 2002 investigation by *60 Minutes* was done. Lawsuits were filed, resulting in some large settlements, but the severe damage continued to harm and kill for four decades. Even with the large settlements, adults and children ended up with less than $10,000 or so to make up for the diseases, disabilities, and deaths they faced each day.

▲ 2000s: Monsanto owns the largest share of GMO products in the world and merges with Pharmacia & Upjohn. It rebrands as an agricultural company and continues to be sued for hundreds of millions of dollars for its toxic products. It files patents for genetically modified farm animals, such as pigs, and files legal actions against small farmers, claiming these small farms infringe on the company's new seed patents. Over 100,000 farmers are said to have been bankrupted by GMO crop failure.

▲ 2005: Monsanto is fined $1,000,000 for bribery in Indonesia. The company authorizes an Indonesian consulting firm to pay $50,000 to a senior environmental official to make him amend the requirements for conducting environmental impact studies before Monsanto can grow GMO crops.

▲ 2014: Monsanto battles state laws to label and regulate GMOs and fight local bans in Hawaii. Monsanto comes under fire for its varietal genetic use restriction technologies (V-GURTs; also known as suicide seeds), which would render first-year planting seeds sterile, forcing farmers to buy their seeds every year instead of saving the best seeds for next season.

This is just a small list of the sins of Monsanto. To list them all would fill a book, and there are many books about Monsanto on the market. Monsanto showed its market control savvy by also becoming a corporate funder of the American Legislative Exchange Council (ALEC), a corporate lobby mill that basically hands state legislators a "wish list" of its patron companies' demands and needs and pays for task forces where corporate lobbyists and special interest representatives vote with elected officials to approve model bills.

Today, Monsanto no longer operates by that name, as it is now part of Bayer's agricultural division, but the damage of its past continues. The biggest issue many had, and continue to have, with Monsanto and companies like it is GMOs.

GMOs

Let's start with a quote from the UN Commission on Genetic Resources for Food and Agriculture. "Cross-fertilizing V-GURT containing crops may cause considerable effects in neighboring crop stands and wild relatives.... The fact that North America, where large stands of GMO varieties are now grown contamination of non-GMO varieties by GMO germplasm has been observed." In addition to literally wiping non-GMO seeds out of existence and destroying, or overtaking, small farms everywhere, GMO machines are dictating public food policy and becoming richer and more powerful, eliminating regulations that would allow consumers to know when they are purchasing and consuming genetically modified products.

What is a GMO? It stands for genetically modified organism and involves the modification and alteration of genetic materials of a particular organism through genetic engineering. The genetic manipulation of an organism's DNA is not natural; it is an engineered method performed in a laboratory. The motives are to create organisms that are resistant to disease, pests, and insects and also to tolerate the use of herbicides, such as Roundup. GMOs also claim to be a method for increasing crop output and food abundance to better supply the world, especially countries with raging poverty.

The majority of corn, soy, sugar beets, and canola crops are GMOs. Salmon is being engineered to grow to maturity twice as fast as salmon in the wild. Pigs and animals raised for food are being manipulated to get fatter faster. Genetically engineered potatoes that resist bruising and contain less of a carcinogenic chemical called acrylamide are being used to make French fries and potato chips, two American junk food staples. Sometimes GMOs backfire. In 2015, Dow Chemical created new corn and soybean crops under the

People stand in front of a crop of transgenic maize. In this case, it is corn with a gene inserted from the bacteria Bacillus thuringiensis *for use in Kenya.*

Enlist brand name that were more immune to herbicides Roundup and 2,4-D, but the collateral damage was the creation of superweeds that would require even more of these herbicides to kill them, thus possibly exposing consumers to more toxins in the end.

Products we wouldn't expect to include GMOs come as a shock to most people. For example, corn and soybeans are the most widely planted GMO crops in the United States and show up in everything from tortilla chips to soy-based meat substitutes. But they also appear in spices and seasonings that contain corn or soy and in soft drink ingredients that are derived from GMO corn, such as corn syrup, colorings, aspartame, and citric acid. It may soon be impossible to avoid GMOs in your diet no matter how hard you try.

Genetically manipulating crops can also help them withstand browning earlier, such as with apples, or surviving under drought conditions. Biotech companies also create new organisms by splicing and modifying the genetic materials of existing organisms: Frankenseeds, Frankenorganisms, Frankenfoods. But they insist there are benefits to these products. These include:

▲ Improving nutritional content

▲ Greater pest resistance and less need for pesticides

▲ Increased crop yield

▲ The ability to give a plant specific beneficial traits

▲ Eases the difficulties of farming crops and helps crops survive stressors better

▲ Lower food production costs for farmers and thus product costs for consumers

In the past, we used selective breeding and crossbreeding of plants and animals to produce better crops and bigger, fatter animals. Traditional breeding methods worked at the time for growing both plants and animals with more desirable traits. But this takes a long time and involves a long process that does not align with today's consumer demands. GMO companies believe that their production takes into account the needs of consumers and speeds up the process for providing them with food and products. Scientists of yesteryear could not make specific and detailed changes to the existing organism to improve up on it as they can today.

 In the absence of long-term studies, their safety is a big unknown, and many consumers don't want to take chances of eating genetically modified foods.

The question is, are they safe? In the absence of long-term studies, their safety is a big unknown, and many consumers don't want to take chances of eating genetically modified foods. There has been an increasing call for labeling GMO foods, even banning them outright, but the companies making these products are wealthy and powerful entities with plenty of backing from politicians. One concern is that genetically altered foods could trigger allergic reactions because they contain so many foreign materials. According to the FDA, GMO companies must test their products for allergens that might be transferred from one organism/food to another during the genetical manipulation, but consumers are forced to take the companies' word that this is being done.

An even bigger concern when manipulating DNA is cancer. Could these products trigger cancer from the genetic mutations?

There is not enough, if any, long-term research to show if consuming food with added genetic materials might alter or affect our own DNA. We do know, however, that these products, which are exposed to glyphosate and other herbicides, pose a huge risk from that and that alone. But what consumer in their right mind would want to eat such Frankenfoods without first knowing what they might do to their health? Sadly, many people have little choice but to eat GMOs as they become more available and widespread. And in poor nations where we send GMO corn and products to feed the starving, those people have even less choice.

Demand for non-GMO foods began skyrocketing in 2013, and many companies took note and told the public they would not engage in using GMO products. Other companies such as General Mills, Stacy's Pita Products, and PepsiCo stated they were creating a line of non-GMO items for concerned consumers. Individual states began passing laws demanding proper labeling. In the United States, GMO products did not have to be labeled. In a Consumer Reports National Research Center Survey in 2015, over 92 percent of consumers wanted GMO products labeled.

In 2015, the Safe and Accurate Food Labeling Act was sponsored by Kansas Republican congressman Mike Pompeo. It was designed to allow companies not to label foods as GMOs. Opponents labeled it "the DARK Act," which stood for "Denying Americans the Right to Know." The DARK Act was full of exemptions and catered to the GMO companies, narrowly defined the GMO products affected, exempted GMO foods where the amount couldn't be "detected," allowed GMO companies to hide their product information behind QR codes, websites, and toll-free numbers, exempted smaller food manufacturers, lacked any enforcement or penalty violation mechanism, and contained a sweeping preemption clause that eliminated all state laws demanding that GMO food and seeds be labeled. The bill died in committee, but the next year, President Barack Obama signed S.744, which, though not as permissive as the DARK Act, was still influenced by it and effectively overturned Vermont's more transparent GMO labeling law.

In 2016 President Barack Obama signed S.744, a bill that was heavily influenced by the Safe and Accurate Food Labeling Act, which made it legal to not label foods as GMOs.

In 2020, the U.S. Department of Agriculture (USDA) began regulating mandating identification of food ingredients that contain GMOs in the National Bioengineered Food Disclosure Standard. This regulation required food processors to inform consumers of the use of GMOs, but although the law went into effect in January 2020, mandatory compliance will not occur until 2022, and the FDA is still struggling with exemptions for products where GMO materials might be lost during processing and refinement stages.

Most packaged foods and many crop foods contain GMOs, and sadly, because of cross-contamination of seeds and processed crop derivatives and inputs, GMOs often find their way into other crops and products. Synthetic biology is responsible for many flavors, sweeteners, oils and fats, vitamins, yeast products, proteins, microbes and enzymes, and hydrolyzed vegetable protein products.

GMOs frighten people because these are novel life forms, made in a lab, often for the sake of increasing profit margins, and completely lack safety standards and testing. The GMO seed revolution has devastated small farming communities by ruining their natural seed supplies and by introducing superweeds and superbugs that become resistant to pesticides and require even more pesticides and ingredients to kill them, increasing the levels of toxicity in the soil and the crops themselves.

But GMO supporters point to the fact that people have been consuming them for over a decade and that there is no evidence of harm. However, just because there is no evidence of harm doesn't mean there isn't harm. In the case of GMOs, there again is little to no actual scientific studies on the long-term effects to establish harm. The question of who would fund any studies undertaken has also been raised. Organizations and lobby groups for Big-Agra? Consumers also wonder if we really know if GMOs are being removed or not used in foods labeled as "non-GMO." Public trust is all but nonexistent when it comes to the possibility of eating plants and meats that have been genetically tampered with.

In November 2020, the Center for Food Safety reported a huge victory when the U.S. District Court for the Northern District of California ruled that the FDA violated core environmental laws in approving genetically engineered (GE) salmon. The court ruled that the FDA ignored the serious environmental consequences of approving GE salmon and the full extent of plans to grow and commercialize the salmon in the United States and around the world violated the National Environmental Policy Act. The court also ruled that the FDA's unilateral decision that GE salmon could have no possible effect on highly endangered, wild Atlantic salmon was wrong, in violation of the Endangered Species Act. The court ordered the FDA to thoroughly analyze any consequences that could result should GE salmon escape into the wild.

"Today's decision is a vital victory for endangered salmon and our oceans," said George Kimbrell, Center for Food Safety legal director and counsel in the case. "Genetically engineered animals create novel risks, and regulators must rigorously analyze them using sound science, not stick their head in the sand as officials did here. In reality, this engineered fish offers nothing but unstudied risks. The absolute last thing our planet needs right now is another human-created crisis like escaped genetically engineered fish running amok."

The plaintiffs were the Institute for Fisheries Resources, Pacific Coast Federation of Fishermen's Associations, Cascadia Wildlands, Center for Biological Diversity, Center for Food Safety, Ecology Action Centre, Food and Water Watch, Friends of the Earth, Friends of Merrymeeting Bay, Golden State Salmon Association, and the Quinault Indian Nation. The Center for Food Safety and Earthjustice represented them.

Deception in a Name

An early promoter of GMOs was the Center for Science in the Public Interest (CSPI), which was founded in 1971 as a supposed food police and consumer advocacy group, which its founders insisted did not take corporate money. However, the CSPI was and is bankrolled by such prominent figures as Bill Gates, the Rockefeller Foundation, Bloomberg Philanthropies (founded by entrepreneur and former New York City mayor Mike Bloomberg), and the American Heart Association. The CSPI has a long history of promoting things like diet soda, GMOs, artificial sweeteners, soy products, trans fats, and fake meats, and it even launched a 1995 campaign to try to get consumers to drink skim milk to reduce their saturated fat intake despite scientific studies showing that full-fat dairy lowers your risk for heart disease and diabetes.

 The CSPI has a long history of promoting things like diet soda, GMOs, artificial sweeteners, soy products, trans fats, and fake meats....

The CSPI, which used an incredibly deceptive name that makes it sound pro-consumer and scientific, worked with the Cornell Alliance for Science, which is a public relations group owned by Gates. In a huge conflict of interest, the head of the CSPI's Biotechnology Project served as a part-time associate director of legal affairs for the Cornell Alliance. the CSPI's history shows it cares less about public health and more about lining its own pockets, and in 2014 it linked with Monsanto to push GMO products as safe, even going so far as to say GMO foods are safer than those that do not contain GMOs. And you can partially blame the CSPI for the United States being the only country that does not clearly label GMO foods, thanks to its heavily intensive pressure and lobbying against labeling.

It is not too much of a stretch to say that anything the CSPI says is safe and good for you is just the opposite, and the CSPI is an example of the deception in food and drug marketing and promotion that leaves the consumers hurting the most. With its links to giant billionaires and their foundations and corporations and their persistent quest to go against scientific research and medical studies, the CSPI can fool the public into thinking it is on the side of the consumer all because of its name.

TOXIC PRODUCTS

"These 10 newborn babies ... were born polluted. If ever we had proof that our nation's pollution laws aren't working, it's reading the list of industrial chemicals in the bodies of babies who have not yet lived outside the womb."

(Representative Louise Slaughter (D-NY), discussing the findings of a study released by the Environmental Working Group. It tested blood from ten umbilical cords and found an average of 287 contaminants, including mercury, fire retardants, pesticides, and the Teflon™ chemical PFOA.)

We live in a toxic world, but the power we hold is awareness, which leads to knowledge. Toxins are in our homes and our offices, and we don't even know we are being exposed. We come into contact with these toxins in four ways:

We ingest them—we eat and drink foods and beverages with toxic chemicals and take toxic pills.

We absorb them—we take them in through contact with the skin or scalp.

We inhale them—we breathe them in via the air, pollution, radon, gases, and the like.

We inject them—we get vaccines with toxic ingredients.

Many of the products we use or consume have at least one toxic element in it, and some have toxic chemical compounds that are truly dangerous to our health. In our fast and easy world—where we want what we want and we want it now—we don't always stop long enough to check labels and ingredients, and when we do, we might ignore the long and complicated names of chemicals or justify them because we are busy and cannot afford to always buy the best. Our health depends on slowing down and facing the truth.

Before we get into specifics, it's important to mention the biggest offenders: asbestos, lead, mold, and radon. These four toxins are present in the places where we live and work and have profound effects on our health and well-being. We feel safe in our homes and comfortable in our workspaces and offices, never knowing the dangers that lurk around us or how we can best avoid and abate them.

Asbestos

Asbestos is a mineral fiber that was widely used in home and building insulation for its ability to withstand heat and act as a fire retardant. It's strong, so it was used in roof shingles, ceiling and floor tiles, paper products, cement products, automobile clutches and brakes, fabrics, packaging, gaskets, and coatings. Most of us are familiar with it being a mainstay of older homes before it was found to be incredibly carcinogenic, especially to the lungs. We have all seen the commercials of the major class-action lawsuits involving people with lung cancer, mesothelioma (rare cancer of the lining of the lung, chest, and heart), and asbestosis (long-term noncancerous lung disease) from the use of asbestos products and exposure to them over the years.

Many homes still have asbestos insulation in walls, attics, and crawlspaces and don't know it until they are told during a home inspection or have an issue that causes them to poke a hole in

Asbestos is a fibrous material that was used for many years in construction as an insulator and fire retardant.

the wall and see it themselves. It is also found in schools and office buildings, especially those built before the early 1980s.

Exposure occurs when the fibers are released into the air from some disturbance of the asbestos materials, say during a demolition, home-improvement project, home repair, remodeling, or office construction or repair. The asbestos is damaged in a manner that allows the fiber particles to become airborne and inhaled. Diseases related to asbestos exposure can take many years to manifest physically, and the risk is higher for anyone who smokes.

Once asbestos is found in a home or office, it cannot and must not be removed by the owner. It requires a hazmat crew of professionals to properly contain and remove all traces of the toxin from the premises. People should consider having someone check their home or office before any kind of remodeling or reconstruction occurs to avoid problems and possible exposure. Asbestos is not banned in the United States, but it has been phased out of many building materials and products because of health concerns and potential lawsuits.

Lead

This potent neurotoxin was once widely used in household paint, water pipes, and plumbing materials until restrictions went into effect in the 1980s. Exposure to lead in paint chips was linked to childhood learning and memory problems; in higher doses, there was damage to the heart, kidneys, and nervous systems of both children and adults. Homes built prior to 1978 may still include lead paint and must be abated when remodeling, along with older plumbing systems that may have lead in the pipes.

In *The Toxin Solution: How Hidden Poisons in the Air, Water, Food, and Products We Use Are Destroying Our Health—And What We Can Do to Fix It*, author Joseph Pizzorno writes about the problems the public faces when trying to get rid of lead. "Instead of studying and routinely eliminating harmful substances before they reach the public, industries, legislators, and regulatory agencies have it set up so that toxins are put out there until a mass outcry gets them removed. The public health measures that got lead out of gasoline and paint were effective: since the ban, levels of lead in human blood have dropped substantially." But lead is still out there in these and other products.

The government and industries have not done such a good job of keeping lead and other toxins out of the products we use, even when they are known to harm children. Pizzorno states, "When it comes to a substance so toxic that it is not safe at any level … how do we wind up allowing it and calling it safe?" This happens because most safe levels of toxins like lead are defined as the levels seen in 95 percent of the normal population, except that the normal population carries such a heavy toxic load and is in poor health and chronic illness, it cannot be used as a healthy baseline.

 The government and industries have not done such a good job of keeping lead and other toxins out of the products we use, even when they are known to harm children.

Even though lead is a naturally occurring element found in small amounts in the crust of Earth and is found in the air, water, and soil, it is toxic to humans and animals. Lead can move from the soil into groundwater and drinking water systems, and natural levels of lead in the soil increase many hundredfold when used in mining, smelting, and other industrial sites.

There are federal and state standards for regulating lead in water, air, soil, food, and consumer products, but it's harder to regulate the exposure to lead paint in homes and offices, especially deteriorating homes and buildings, such as old apartment complexes. But lead is also released in the making of stained glass.

Lead exposure is especially dangerous to pregnant women because the developing fetus can be negatively affected. Babies can be born prematurely, often with damaged brains, kidneys, and nervous systems, and the mother is at a higher risk of miscarriage. Children six years of age and younger are highly susceptible to lead's effects on their brains and nervous systems. The toxin accumulates in the body over long periods of exposure, putting those who live and work in old buildings at a high risk for disease that even includes high blood pressure, decreased kidney function, and infertility. Ingesting lead can cause seizures, coma, and death.

Mold

Mold is a type of fungus that has existed for millions of years on Earth and thrives on moisture. Mold can reproduce through lightweight spores that are airborne. The small organisms can be black, white, orange, green, or purple and are found indoors and outdoors. They are usually harmless in small amounts of exposure and we are exposed to various molds every day, but inside the home, they can proliferate and release more spores that are inhaled and become a health issue, sometimes a serious one.

In nature, it is the job of mold to break down dead organic matter such as leaves and dead trees. Indoors, mold is commonly found around leaks in roofs, doors, windows, basements, heaters, air conditioning units, and high-moisture areas, but it also grows on paper products, carpet, drywall, wallpaper, in plants, dust, fabric, upholstery, ceiling tiles, cardboard, and anywhere there has been flooding. It can spread through dust, the air, and contact by humans and pets, attaching itself to shoes and clothing. Spores are deposited indoors and

Mold is a treacherous contaminant once it gets into your home. It can hide for years, and the spores can get in the air and cause respiratory issues, infections, and even internal hemorrhaging in worst-case scenarios.

thrive, becoming large infestations if not detected and dealt with. They may or may not smell but can always be seen.

Some of the health issues caused by exposure to mold include asthma, wheezing, sore throat, eye burns, allergies, skin reactions, cough, upper respiratory infections, and immune-related conditions. The World Health Organization (WHO) in 2009 issued studies showing links between early mold exposure in infants and the development of asthma and respiratory allergies. Other studies showed mold caused idiopathic pulmonary hemorrhage among infants, memory loss, lethargy, and fungal infections. Individuals such as seniors, children, and the immunocompromised were most at risk of the toxic effects of molds.

Certain molds are "toxigenic," which means they produce toxins called "mycotoxins." Like common household molds, they can be toxic with long-term exposure. Often, people don't know they have mold until they do some home improvements, move furniture, or have a storm with some flooding. This means mold can be growing and releasing spores for years before anything is ever detected, and those spores can cause illnesses that increase in symptoms over time, not to mention you might not first think of mold as the cause of your respiratory or breathing problems and wait to even mention it to your doctor.

It's almost impossible to get rid of all mold spores inside the home because spores land on surfaces or in dust and hide in corners despite a good cleaning. However, they won't grow unless there is moisture present, so the best you can do is control the moisture levels in your home and keep everything clean and dry, and if you have mold, clean it up and be vigilant so that it doesn't return. If your heating, ventilation, and air conditioning (HVAC) system is contaminated with mold, you can consult the Environmental Protection Agency's (EPA) website for help cleaning it or call a professional. Like asbestos, you don't want to assume you got all the mold out of your house just because you can't see the individual spores just waiting to spread.

During humid weather, you can use a dehumidifier in the home, too, and keep indoor humidity between 30 and 50 percent ideally. Don't forget out-of-the-way places mold might proliferate like rain gutters, mudrooms, laundry rooms, above kitchen cabinets, under and behind refrigerators and iceboxes, and under sinks in the bathroom and kitchen.

Radon

Radon is a radioactive gas that is found in trace amounts in natural stone, which can also include uranium. As these elements decay, they produce radon gas, which is colorless and odorless, so you won't be aware of the potential damage it is doing. It has been linked to lung cancer and respiratory illnesses, but it can be dissipated with good ventilation. Radon tends to accumulate in small, unventilated areas like attics, basements, pantries, and crawlspaces, especially during cold weather months when windows remain closed. Radon is also emitted from the soil homes are built upon.

The EPA estimates that there are thousands of cancer deaths each year in the United States from radon exposure. Nearly one in fifteen homes in the United States are estimated to have elevated levels of radon and in 2005, U.S. surgeon general Richard Carmona warned that radon was the second leading cause of lung cancer.

There are companies that will do full radon tests in homes, and you can now buy home tests online that

Homeowners can now purchase digital radon detectors to check for dangerous levels of the gas.

will help you find the levels of exposure. It's suggested that in homes, all floors below the third floor should be tested. Mitigating the problem might require getting rid of the suspected product or increasing the amount of ventilation in the area. Luckily, there are many homes now being built with radon-resistant construction features. Each state also has a radon office that can help pinpoint local levels and the best way to test your home.

Poisonous Chemicals

The list of products we use daily that contain one or more chemicals or toxins is overwhelming, so this list is not comprehensive. It is in alphabetical order for convenience and presents the reader with some semblance of the scope of toxicity we deal with every time we buy, use, or consume a product or a service.

There are six classes or families of chemicals in consumer products that negatively affect human health. These are:

▲ Highly fluorinated products

▲ Antimicrobials

▲ Bisphenols and phthalates

▲ Solvents (not all)

▲ Flame retardants

▲ Specific metals

Many products contain combinations of the above. Buyer beware.

Poisons on the Airlines

Inside the cabin of an airplane, there are high levels of toxins that cause a cluster of symptoms, including headaches, respiratory distress, dizziness, and neurological issues, all indicative of aerotoxic syndrome. One of the biggest culprits is tricresyl phosphate (TCP), found in a number of popular airline cabins where passengers were exposed to this toxin during flights. According to Sarah Mackenzie Ross, clinical neuropsychologist at University College London, over 200,000 passengers may suffer from aerotoxic syndrome each year, and a 2008 investigation by the *Daily Telegraph* uncovered hundreds of complaints filed by British pilots suffering the above symptoms.

TCP and other toxins are sucked into the cabin as a consequence of engine design, which uses a "bleed air" system that removes air from the engine's compressor, cools it, then funnels it back into the cabin after filtration from bacteria and viruses. But there is no filtration of toxic particles or fumes, forcing many airlines, but not all, to come up with an alternative to the bleed air system for providing air to cabin passengers.

 TCP and other toxins are sucked into the cabin as a consequence of engine design, which uses a "bleed air" system that removes air from the engine's compressor....

Imagine being a flight attendant and finding out your uniform is toxic to your health. A number of lawsuits have been filed against Lands' End clothing company for producing flight attendant uniforms to Delta airlines that caused respiratory distress, skin reactions and blistering, hair loss, nosebleeds, migraines, and other autoimmune conditions. In May 2018, approximately 68,000 Delta employees were given such uniforms consisting of dresses, shirts, jackets, pants, tops, and sweaters that contained various chemical additives and finishes meant to keep the clothing from wrinkling and losing its stretchiness. Another additive treated the uniforms to be self-deodorizing and stain resistant.

By August 2019, over 1,900 employees had reported adverse health conditions and the lawsuits mounted, especially after the National Institute for Occupational Safety and Health released a health hazard evaluation in June of that year stating, "It is possible that textile chemicals in the uniforms or the physical irritant properties of the uniform fabrics have caused skin symptoms among Delta employees."

The plaintiffs included in their lawsuits tests of the toxic uniforms that found six heavy metals and chemicals, including bromine, mercury, and formaldehyde in levels in excess of industry safety standards. In 2018, Alaska Airlines was hit with similar lawsuits after employees wore new uniforms and reported rashes, eye irritation, blistering, hives, and scaly patches on the skin and was backed up by a Harvard University study of cabin air quality that revealed data from 684 flight attendants before and after the uniforms were issued. The study found that respiratory, allergic, and dermatological problems rose after the uniforms were issued.

Baby Food

It can be terrifying to think of feeding our children toxins. We tend to trust that anything directed toward babies and infants must be healthy, or it would never be permitted on store shelves by government oversight committees and regulators. Sadly, this is not the case. In October 2017, the Clean Label Project, a nonprofit group that advocates for transparent labeling, did a test on baby food, formulas, and toddler drinks and snacks all purchased during a five-month period.

A 2017 study found that 80 percent of baby formula contained arsenic.

The results were stunning. The study found that 80 percent of infant formula tested positive for arsenic. It also found that 65 percent of the other products, which included popular brands and emerging brands, tested positive for arsenic, 36 percent for lead, 58 percent for cadmium, and 10 percent for acrylamide. These chemicals have all been linked to negative issues with fine motor skills and cognition.

As we've seen earlier, arsenic is associated with developmental defects, cardiovascular disease, and neurotoxicity, according to the WHO. One of the worst offenders in many baby foods is rice. The U.S. Department of Agriculture (USDA) imposed a limit of 100 parts per billion of arsenic in infant rice, but there is little enforcement of that limit. Rice can absorb arsenic through the soil it grows in, making it all but impossible to remove the toxin.

Lead has been found in baby food products before, and low levels of lead in the blood can result in lower IQs, stunted growth, anemia, and behavioral issues. Of the 2,164 baby food samples tested in the Clean Label Project, 20 percent contained lead.

In October 2017, CNN did a spotlight report on toxins in baby food and referenced a study called "Healthy Babies, Bright Futures: Lowering the Levels: A Healthy Baby Foods Initiative," in which the researchers found of 168 baby foods sold in the United States, 95 percent contained lead, 73 percent contained arsenic, 75 percent contained cadmium, 32 percent contained mercury, and most contained pesticides and mycotoxins. One-fourth of the baby foods contained all four of the above heavy metals! Again, rice cereal and rice products were the biggest offenders.

Sadly, even organic rice foods contain arsenic, and whole grains are even worse because the heavy metals are more absorbed into the outer layer of the grains. Parents have been told to focus on whole foods and organic, pasture-raised, and nutrient-rich foods, and, if possible, to make their own baby food from organic veggies, fruits, and meats.

The "Healthy Babies" study went on to state, "These four harmful metals are found in all food—not just baby food. They occur naturally or from pollution in the environment. Crops absorb them from soil and water, and they are even found in organic food. Their presence in baby food raises unique concern because babies are more sensitive to the toxic impacts."

Plastics are another problem, with a research study published in the October 2020 issue of *Nature* suggesting that heating up plastic baby bottles releases tons of microplastic particles into the milk or formula, exposing the baby to hormone-disrupting chemicals and increasing risks of heart disease, diabetes, and cancer. Most baby bottles are made with a common plastic called polypropylene, and in the study, researchers found that after heating up baby formula in a number of plastic bottles, they detected a range of 1,300,000 plastic particles per liter of formula to a whopping 16,200,000 particles per liter.

On average, the study found that an infant consumes about 1,600,000 particles of microplastic particles per day, 2,600 times the amount the average adult consumes on a particular day. The study also warned that it's how the baby bottles are heated that also adds to the problem. The WHO recommends sterilizing all baby bottles before feeding, but in the course of doing so, the level of microplastics in the formula increases due to hot liquid releasing more plastic from the bottle than cold liquid. Plastic kettles should also never be used for the same reason. Reheating the same bottle also increases the amount of plastic released, especially doing it in the microwave. Even shaking the bottle releases plastics.

Ideally, glass bottles are the way to go.

Toxins are everywhere, but there are steps we can take to minimize their impact on our health, and we owe it to our babies to get them started on the best track to health as possible.

Bedding

There is nothing like a new pillow or mattress for improving sleep. But bedding comes with a host of toxins that include volatile organic compounds (VOCs) and flame retardants (see below section) that are inhaled not just during sleep but anytime you are in the bedroom. These chemicals irritate the lungs, cause cancer, and trigger allergies, even rashes to the skin. They can also dampen your immune system and disrupt hormone production. This is especially frightening because you are close to these products during sleep when your body is at rest and more susceptible to the damage of the toxins.

Pillows and mattresses outgas chemicals, which are inhaled during sleep and can cause headaches, vision problems, dizziness, memory impairment, kidney and liver damage, and

Those new bedsheets and pillows you bought might seem fresh and clean, but they could also be contaminated with flame retardants and VOCs.

central nervous system damage. Because your pillow comes into contact with the delicate skin of your face, you can also experience eczema, atopic dermatitis, and eye problems.

The solution isn't just to clean new bedding but also to cover mattresses and pillows with allergy-preventing pillowcases that prevent exposure not just to toxins but also to dust mites and dander. Pillows that are made of synthetic foam are the worst, so look for those made from natural wool, organic cotton, or plant-based filling, and the casing should also be organic. Mattresses should contain at least 95 percent organic content and have a low-VOC certification. Make sure they are also flame retardant and PVC-free, too. They may cost more, but it will pay off in better health. This also applies to quilts, comforters, blankets, and sheet sets.

Carpets and Flooring

Though many new homes opt for tile and wood flooring, others contain wall-to-wall carpeting or carpeting in specific rooms. Carpet is a trap for toxins and allergens, not to mention mold, and the toxins that are burrowed into the carpet come back up in the form of VOCs and chemical compounds that are then inhaled or absorbed by the skin. Carpet fibers are treated with chemicals meant to prevent staining, strengthen the fibers, and make the carpet integrity last longer, but they in turn damage human and animal life. Children and animals are most at risk, as they often play on the carpeting and even lay down on it for hours at a time.

Many types of laminate flooring contain formaldehyde, which can release dangerous fumes into your home. Some flooring is safe, but a floor with high levels of this chemical can release fumes for up to ten years.

Carpet chemicals include triclosan, a thyroid and reproductive system disruptor; formaldehyde, a carcinogen; permethrin, a pesticide and neurotoxin that damages the central nervous system; and polybrominated diphenyl ethers (PBDEs), which disrupt thyroid function, cause developmental delays in children, and cause infertility. These chemicals may make the carpet softer, cleaner, less receptive to fire, and last longer, but as you can see, they have harsh influences on our health. Even carpet padding contains plastics made of petroleum and other chemicals, and it may come as a shocking surprise to know what the carpet we throw away does to the environment. Most carpets end up in local landfills, to the tune of almost 1,800,000 tons, and they can last over 20,000 years.

Who has time to vacuum several times a week with an expensive HEPA filter vacuum, considering that is what it would take to keep these toxins under control? Wool carpeting is less of a toxin trap, and area rugs are better because they cover a smaller amount of floor space, but they still should be avoided. Even worse, people add stain treatments and waterproofing products (such as Scotchgard) to carpeting, which compounds the number of toxins they contain.

Wood flooring and tile have their own issues, with VOCs and treatments that protect them from moisture and staining. One widely used alternative to carpeting is vinyl flooring, which contains PVC and releases dangerous phthalates inside the homes, mainly accumulating in dust and later ingested by everyone, including pets. The National Resources Defense Council (NRDC) found that all homes contain levels of diethyelhexyl phthalate (DEHP) contaminants, and the average levels exceed EPA contamination regulations.

Vinyl flooring also contains vinyl chloride, a known carcinogen and VOC. Other chemicals in this type of flooring include dioxins, ethylene dichloride, and mercury, released by the PVC, and these chemicals are labeled highly persistent and bio-accumulative, which means that once they enter the environment, they can remain there for decades and are hard to get rid of.

Older flooring and flooring made in China are notorious for containing formaldehyde and should be avoided. Better flooring choices that are affordable are linoleum or tile made of ceramic, porcelain, and glass, although you must make sure grout, adhesives, and sealants are made of nontoxic materials. Cork and bamboo flooring are natural options, too. Hardwood is considered the best option but can be expensive and harder to maintain, which might mean using more toxic wood floor cleaners over the long haul, and wood is not a great option for places in the home where there is a lot of moisture, such as kitchens and bathrooms.

The Environmental Working Group (EWG) recommends looking for a Green Seal 11 certification on flooring products.

Clothing

Textiles, including clothing, are big polluters, and the textile industry contributes about 20 percent of the global water pollution. Fashion has its downside, and this includes the presence of over 43,000,000 tons of chemicals per year that are used in the production of textile products. This does not include any pesticides that are used on the cotton crops that make their way to our wardrobes and closets.

> Fashion has its downside, and this includes the presence of over 43,000,000 tons of chemicals per year that are used in the production of textile products.

The clothes we wear can contain flame retardants (see below), formaldehyde (used to take out wrinkles and stop shrinkage), azo dyes (for color), the big heavy metal three (cadmium, lead, and mercury, found in dyes and leather tanning agents), and perfluorinated compounds (PFCs, used to make waterproof and stainproof clothing). All of these are carcinogens and cause plenty of other troubles as well, including liver and hormone problems.

Because we wear clothing close to the skin, we run even greater risks of absorption of these chemicals, especially if we have cuts or scrapes that allow them into the skin and bloodstream. There are organic, chemical-free lines of clothing, such as those made from hemp, but most people prefer the cheaper and easier-to-find clothes they can get at any big-box store or department store. Used clothing is great if you're on a budget, but keep in mind many thrift stores and consignment shops add perfumes to clothes to make them smell fresh and new, and often, these fragrances last after several washings, risking more exposure into the lungs and airways.

Coffee

Coffee is one of the most popular beverages in the world. Depending on how it is processed, it can contain different levels of harmful elements such as toxins and molds. On the good side, it contains polyphenols and antioxidants.

Coffee is one of the most chemically treated crops, too, but luckily, roasting the beans removes most of the toxins, which come from mycotoxins that are produced by molds. Studies by the European Food Safety Authority, the Food and Agriculture Organization, and the World Health Organization show that exposure to these toxins are at only about 2 to 3% of the dosage considered to be unsafe, even if you drink four cups daily.

What is more concerning is water pollution that results from coffee processing. There are two methods in which coffee beans are prepared: wet (or washed) processing and dry (or natural) processing. In the former, the cherry (the fruit around the bean) is stripped off the bean, which is thoroughly washed to remove any pulp, then dried; in the natural method, the fruit and bean are dried together and then the fruit is stripped off. The wet-processing method takes about 10 liters of water per kilogram of coffee, and the dry method takes about one liter.

Wet-processed coffee is often considered superior by coffee afficionados because leaving the fruit on during the drying of the bean means the coffee absorbs more flavors and has a richer taste, but this method alarms environmentalists because it uses so much water and also contaminates it. According to a *Specialty Coffee Association News* article by Carlos H. J. Brando:

> The growing concern for the environment in the last quarter of the last century led to questioning the use and contamination of so much water—often 10,000 m^3 per ton of green coffee—in the wet processing method. The pollution load in the wastewater from the wet milling of coffee can be 30 to 40 times greater than the one found in urban sewage!

Decaf coffee is processed with a variety of chemicals that have been labeled as carcinogenic. There are three methods for removing caffeine from coffee. One is with liquid carbon dioxide.

Another uses a chemical solvent, and the last uses water. Water-processed decaf is the best option, but all decaf contains some levels of aflatoxin and ochratoxin, both of which are linked to cell mutation and organ toxicity in humans. Solvent-method decaf coffee also contains synthetic chemicals such as ethyl acetate and methylene chloride, the latter of which is used in adhesives, paints, and pharmaceuticals.

Paper coffee filters that are white are bleached and when exposed to hot water can release contaminants into your coffee. Paper bleaching is a resource-intensive and polluting process. As the coffee is filtered, the hot water serves as an extracting solvent for chemicals in the paper, the most

Dry-processed (also called natural-processed coffee involves drying the bean with the cherry prior to depulping. Wet-processed coffees (pictured) have the cherry removed first before washing the bean. Wet-processed coffee uses ten times more water than dry.

concerning being chlorinated dioxins. These dioxins can increase the risk of cancer. There are unbleached paper filters on the market, but it is better to use a mesh wire filter that you don't have to replace every time you make coffee; it's a lot more friendly to the environment, too.

Cookware

It's one thing to be aware of the toxins in the foods we eat. It's another to consider toxins in the cookware we use to cook many of those foods. Heavy metals and contaminants exist in various amounts in much of what we use, and often, the release of these toxins is directly related to damaging chemicals leaching into the foods from overheating as well as the actual chemicals themselves.

The most notorious toxic cookware, Teflon™, was responsible for perfluoroalkyl substances (PFAS), now found in the bloodstreams of almost everyone in the country, according to "Is Your Cookware Killing You?" from Craig Stellpflug for the May 2012 issue of *Natural News*. The perfluorooctanoic acid (PFOA), which has been linked to cancer, thyroid disease, and infertility, among other health ailments, ends up in food when the surface of the pan or pot breaks down from high heat, emitting fumes that can also cause "polymer fume fever," similar to the flu.

Europe banned the use of PFAS in 2008, but the United States didn't ban the toxin until 2014, and this was only after decades of exposure that continues to result in illness and legal action.

Teflon™ is so attractive to consumers because it is nonstick and easy to use, but PFAS are known as "forever chemicals" because they never go away, even in the blood and organs of human beings. Studies done by the Centers for Disease Control and Prevention (CDC) showed that 98 percent of Americans have PFAS in their blood, and even at low doses, such as the amount of PFAS found in our drinking water, they increase the risks of deadly disease. Sadly, the EPA and the Department of Defense downplayed the risks of the chemicals used to develop Teflon™ and another popular product, Scotchgard. Since Teflon™ is also found in waterproof clothing, furniture, self-cleaning ovens, pizza boxes, and microwave popcorn bags, the presence of PFAS are everywhere, not just on our pots and pans.

 In 2015, manufacturers were ordered to stop making products with PFAS, and today's Teflon™ is PFAS-free.

In 2015, manufacturers were ordered to stop making products with PFAS, and today's Teflon™ is PFAS-free. However, Teflon™ is found in antistain coatings used on furniture and carpeting, which many older homes still have.

If you are using aluminum cookware, a common and affordable choice, the heavy metal itself can be absorbed into food during the cooking process. Aluminum in excess is associated with a host of illnesses including cancer, Alzheimer's, Parkinson's disease, and autism. Cooper cookware is better at conducting heat, and it releases a tiny bit of copper into food, which is fine, but it also releases the highly allergenic heavy metal nickel, which is found in the coating.

Cast-iron cookware can leach iron into food, and it changes the enzymes in food. If used too often, it increases the iron intake to possibly dangerous levels. It can take high heat, though, and no toxic fumes are released while cooking. Ceramic, enamel, and glass cookware must be checked for the presence of lead, which is often added to give the product a uniform color. Stainless steel cookware poses the problem of leached metals like iron, chromium, and nickel into foods, which can cause health issues. Titanium cookware seems to be fairly safe and doesn't allow chemicals to be leached into food, but a high-quality cookware set can be pricey (although worth it for your health).

Cooking with high heat releases not just carcinogenic chemicals from cookware but often from the food itself, and it also compromises the nutritional value of the food. This is especially true of cooking meat on high temperatures or grilling, which releases carcinogens. If possible, a good set of titanium cookware and keeping cooking temperatures below 200 degrees can be best for your health.

While we're on the subject of cooking, so many people save their leftovers in aluminum foil, not realizing aluminum is itself a neurotoxin and when used to heat and cook food can increase the risk of neurotoxic effects and diseases. Using foil too often also increases the risk of bone disease and promotes pulmonary fibrosis when used to prepare, store, and cook food. This lung disease is more likely to occur from people exposed over the long term to aluminum oxides, but over a long period of time, even everyday kitchen use can pose a threat.

The foil interacts with food by leaching aluminum compounds. Cooking acidic foods like lemon juice, limes, or tomatoes cause the most leaching, made even worse when certain spices are added. The aluminum in the foil is not inert and reacts to different foods by leaching a portion of its compounds out into the food, and when ingested over time, it causes a buildup of aluminum in the blood, muscles, and organs.

For baking and for storing food more safely, use glass containers, as glass is inert.

Nylon cooking utensils are inexpensive, but many contain primary aromatic amines, PFAs, that have been linked to cancer. They are heat resistant but not heatproof, so they can melt upon contact with superhot surfaces. When they begin to wear out, they can flake into food. Some black plastic utensils are made from recycled plastics from electronics equipment and could contain PVC, heavy metals, and flame retardants. Best bets are wood or bamboo utensils.

Cosmetics and Skin Creams

Billions of dollars' worth of cosmetics are sold each year, but many users have no clue as to how these products can harbor all kinds of toxins and even bacteria-resistant superbugs. In the January 4, 2020, analysis by Dr. Joseph Mercola, "Is Your Makeup Contaminated with Superbugs?", an analysis of over 460 donated

A 2019 United Kingdom study found that 79 to 90 percent of makeup products were contaminated with bacteria such as E. coli and Staphylococcus aureus.

makeup products were tested and showed that 90 percent were contaminated with bacteria, including fecal contamination. Some of the worst products included makeup sponges, the vast majority of which were not properly cleaned or ever cleaned at all. Some showed signs of being dropped on the floor and reused! Sponges, used for blending cosmetics, are often put away damp after being used, which allows the bacteria to multiply.

Some of the bacteria found included the antibiotic-resistant superbug, which can trigger deadly stomach issues, diarrhea, and even death. Commonly found were *Staphylococcus aureus*, *E. coli*, and *Citrobacter freundii* as well as fungi and Enterobacteriaceae. All of these can trigger illnesses when in contact with the eyes, nose, or mouth or any breaks in the skin. This is made worse when products are used past their expiration dates. The Food and Drug Administration (FDA) does not require cosmetics to post a shelf life or printed expiration dates on the product packaging, making matters worse for the consumer. Since the industry is also self-regulated, it's impossible for the consumer to know where the product was made or tested or if any restrictions were placed on ingredients.

Personal-care products, which include cosmetics, are manufactured with toxins. Europe banned these products, but there are no restrictions in the United States. The average woman uses twelve different cosmetic products, and combined, they can contain close to 170 different chemicals including parabens, triclosan, phthalates, phenols, and BP-3. Even cosmetics labeled paraben-free or phthalate-free, according to the analysis, contained increases in other chemicals. An evaluation of approximately fifty makeup products found heavy metal contamination in almost all of them, including lead, beryllium, thallium, cadmium, and arsenic. Cosmetic manufacturers can use almost any raw material to make their products without approval from the FDA.

Because cosmetics can be pricey, many users don't throw them away often, and they use the same applicators and sponges over and over. Some suggest that makeup used on or near the eyes and mouth, such as mascaras and lipsticks, should be thrown away every six months and that applicators and sponges be thoroughly cleaned after use or disposed of for a new one. All too often, people forget when they purchased something and are reluctant to part with it because of the cost.

Parabens, which are also found in face creams, aerosol deodorants, hair products, and toothpastes, mimic human estrogen and can cause havoc to a woman's endocrinal system. They also have been shown to accumulate in breast cancer tumors, even in people who don't use the products regularly.

 Congress has attempted to pass bills that require manufacturers to register products and ingredients and submit them for FDA review, but none have managed to pass....

Congress has attempted to pass bills that require manufacturers to register products and ingredients and submit them for FDA review, but none have managed to pass, most likely due to the powerful lobbying efforts of the beauty industry that appears to be only skin deep and care little for the health of their consumers.

Skin creams, lotions, and serums contain chemicals that have never been tested for safety or have already been found to cause harm, thanks to the law that protects the safety of personal products being outdated. In fact, it hasn't been revised since 1938!

Some of the chemicals sound natural, such as alpha and beta hydroxy acids, which exfoliate the outer layer of skin to reveal the newer layer. These products instead reveal sensitive skin to potential sun damage. Skin lighteners contain hydroquinone, which has been linked to respiratory illness, cancer, eye damage, and a skin condition called ochronosis, which causes skin to thicken and turn blue-gray in color.

Facial cleansers and moisturizers often contain polyethylene glycol (PEG), which may be contaminated with carcinogens that allow the product to penetrate the skin and therefore make the toxic chemicals more readily absorb into the skin. Even skin products that are labeled organic or natural can contain toxic chemicals and can claim they are organic or natural if they contain just one ingredient that fits the bill. If the product is not independently verified as USDA Certified Organic, it can contain a lot more chemicals than it is letting on. This is called greenwashing. The USDA Certified Organic designation requires that products contain 95 percent or more organic ingredients, with the remainder being on a list of safe ingredients.

Men are not immune. Shaving gels and creams sometimes contain polytetrafluoroethylene (PTFE), also known as Teflon™. This helps give gels a slippery texture. It sometimes shows up on the product label as polyfluoroalkyls or perfluoroalkyls, so it's best to avoid anything with the word *fluoro* in it. These chemicals cause cancer, immune and thyroid defects, and liver damage and have been linked to birth defects.

Colognes and perfumes are some of the most toxic beauty products around and are especially problematic because the boxes or containers they come in don't list any ingredients, leaving the consumer in the dark. Fragrances contain combinations of synthetic chemicals to achieve the proper scent, which are linked to cancer, respiratory illnesses, reproductive issues, skin rashes, and allergic reactions.

One of the most widely used anti-aging treatments is itself a toxin. Botulinum toxin, best known as Botox®, is used in injection form to reduce the appearance of facial lines and wrinkles. The particular toxin used is onabotulinumtoxinA, and its purpose is to temporarily paralyze or freeze a muscle so that it cannot move. Other products that use this toxin include Dysport, Xeomin, and Myobloc.

Botox® injections are not just meant to make you look younger by getting rid of frown lines; they are also used to treat neck spasms, excess sweating, an overactive bladder, lazy eye, eye twitching, and chronic migraines. The toxin has also been used as

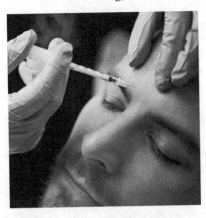

a treatment for cerebral palsy (CP) by paralyzing stronger muscles to allow weaker muscles to do the legwork (no pun intended). CP can cause limbs to pull into the contract, causing spasticity, and Botox® relaxes the muscles so they can be worked with in physical therapy.

There have not been long-term studies on the injection of Botox® for such therapies, but they do work. Still, they come with risks and complications, including pain, swelling, bruising at the injection site, headache, flu-like symptoms, droopy eye, crooked smile, drooling, eye dryness, excessive tearing, muscle weakness, vision problems,

Botox® injections contain the Botulinum toxin, which temporarily paralyzes muscles.

breathing problems, loss of bladder control, difficulty swallowing, and trouble speaking. There is also the risk the Botox® will travel somewhere it is not intended to, which means if you plan to get injections, you must find a professional or doctor who is experienced with the placement of the needles.

Those who should avoid Botox® are pregnant and lactating women and anyone allergic to cow's milk protein. Anything injected into the body has the potential to drift from the injection site and end up near or in a vital organ. Unless it's a part of a medical procedure, there are better and safer choices for looking younger.

Cleaning Products

Household cleaners are notorious for the number of toxic chemicals they contain. Now, we understand why our parents kept cleaning agents locked away or on high shelves. They even smell like caustic chemicals. These include everything from bathroom toilet cleaners to countertop cleaners to floor cleaners to laundry detergents to rust removers.

A large majority have been found in various consumer-advocate tests to contain ingredients that cause respiratory problems and cancer, even though there are many alternative cleaning choices on store shelves that are much less toxic.

Cleaning agents include stabilizers, which allow the formula to have a longer shelf life but are linked to eye, nose, throat, skin, and lung irritation. Dioxane causes respiratory problems, hormone disruptions, kidney disease, and cancer and impairs the brain and immune system. Benzene causes cancer of the blood. Sodium lauryl sulfate is toxic to human organs and causes cancer and reproductive and neurological problems. Added fragrances contain their own laundry list of toxins. Finally, bleach products contain a known skin, eye, nose, and throat irritant—bleach. And when combined with water, it can cause organic compounds that are toxic and dangerous to the lungs, kidneys, and liver.

The worst products are toilet cleaners, oven cleaners, tarnish removers, floor cleaners, and ammonia-based products, which all release high levels of hydrochloric acid, phosphoric acid, sodium, potassium hydroxide, and ethanolamines. These cause lung inflammation and potentially cancer. Using these products often comes with a bout of dizziness, wooziness, and trouble breathing, especially without proper air ventilation. The thing is, these products smell so powerfully chemical, you cannot help but

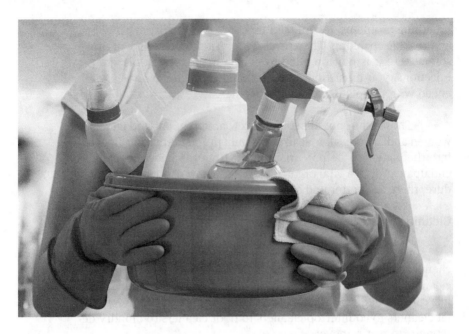

Many common household cleaners can prove toxic when not handled correctly, or even when they are!

wonder what those chemicals are doing to your body when inhaled and absorbed.

Long-term exposure to cleaning products and agents can damage lungs to the point where they resemble the lungs of a heavy smoker of ten to twenty years.

In recent years, there has been a growing concern about the role of trichloroethylene (TCE) in causing a rise in Parkinson's disease, especially in the United States (the chemical is banned in the European Union). TCE is a chemical used in degreasing cleaners. It is used in industry but also in household products such as carpet cleaners and shoe polish. According to an article in *The Guardian,* a 2008 study published in *Annals of Neurology* indicated a link between exposure to TCE and contracting Parkinson's. It is also a carcinogen known to increase the risk of cancers of the liver, cervix, lymph nodes, breast tissue, and biliary passages, and it can also cause cardiac defects in fetuses.

According to the *Guardian* article, about 30 percent of U.S. groundwater is contaminated with TCE, with Santa Clara, California, being a huge hot spot. Residents there are surrounded by no fewer than 23 Superfund sites. The EPA did a study in 2012 and

Phthalates

Pronounced *THA-lates*, these plastic softeners are found in so many products we use daily, including shower curtains, shampoo and conditioner bottles, perfumes, rubber boots and raincoats, car dashboards, even baby teething rings. They are linked to endocrinal system ailments and are considered hormone disrupters, but they are also linked to lower sperm counts, sexual disfunction, and breast and prostate cancers. They can also cause genetic mutations. They are illegal in Europe, where stronger regulations are in place due to nationalized health care, but are perfectly legal in the United States. The FDA does not regulate or act as a watchdog over products, including phthalates, which is why they can even be used in baby products.

2013 revealing that residents there were literally inhaling TCE vapors that were rising from the ground beneath their feet. Another hot spot was Camp Lejeune in North Carolina, where from the 1950s through the 1980s, pregnant wives of Marine Corps personnel were exposed to TCE levels 3,400 times higher than what is considered safe. The result was a crisis of stillborns. In fact, soil and groundwater pollution is a chronic problem at U.S. military bases across the country, as there seems to be little concern about dumping fuel and chemicals like TCE near the homes of military personnel.

TCE dumping is still a problem in the United States. In 2017, American industries released over two million pounds of TCEs into the environment. There is some glimmer of hope, however. In 2020, Minnesota became the first U.S. state to ban TCE dumping, and New York state is now considering a similar ban.

Dryer Sheets and Fabric Softeners

Even trying to dry your clothes and get rid of static can be toxic. Dryer sheets and fabric softeners contain a number of toxins that have not been tested in any specific combinations, and because they get in or on your clothes, residue eventually makes its way to you, causing allergic and respiratory reactions, nervous system disorders, vomiting, nausea, dizziness, headaches, syncope, and more.

The chemical cocktail includes:

▲ Alpha Terpineol, Pentane, and Linalool: potentially air-borne substances that can cause illness when inhaled.

▲ Benzyl Acetate: a potential carcinogen, which can cause pancreatic cancer.

▲ Chloroform: a known carcinogen and neurotoxin that when released into the air and heated can cause loss of consciousness, vomiting, and dizziness.

▲ Ethanol: on the EPA's Hazardous Waste list.

▲ Ethyl Acetate: on the EPA's Hazardous Waste list.

▲ Fragrance: synthetic chemical compounds that can cause allergic reactions and respiratory distress.

Dust and Soot

All the chemicals found in various household products eventually make their way into the dust that gathers on your floors and furniture. This includes pesticides, flame retardants, cleaning chemicals, microfibers, and dust from electronics, which contain a high content of flame retardant. It's not enough to do a quick dusting now and then because this kicks up more toxins back into the air and then back to the ground. To really rid your home of toxic dust, you need a vacuum or air cleaner with a HEPA filter and a good wet mop of the floors first, then a microfiber or damp cloth clean sweep of furniture (without first spraying even more toxic cleaner onto the cloth, please!).

High-Efficiency Particulate Air (HEPA) filters prevent particles of 0.3 microns in size from passing through 99.97 percent of the time.

High-quality HEPA filters cut down not only on the amount of toxins but also pet dander and dust mites, so it's worth the investment.

Chimneys give off soot that is filled with microparticles, including carbon and the incomplete combustion of coal, oil, wood, and other products. Soot is filled with chemicals, even heavy metals and acids, and the particles can be so small that some air cleaners won't pick them up. It becomes inhaled or absorbed through the skin and causes dizziness, nausea, respiratory issues, and, with prolonged exposure, could contribute to lung disease.

The fine, black powder that makes up soot is filled with lead, arsenic, chromium, and cadmium. Obviously, it's critical to keep the chimney clean, but existing soot may go unnoticed and, when kicked up by a breeze down the chimney, become an airborne toxin.

Wood dust is another offender, whether you're sawing wood for a home project or walking through a bar with a sawdust floor. Exposure to sawdust can be very problematic for anyone with asthma, respiratory illness, and cardiovascular disease. People who work with sawdust wear safety goggles and respirators to avoid these potential toxins that have been linked to nasopharyngeal carcinoma and lung cancer.

Flame Retardants

Their purpose is to act as fire retardants and possibly help save your life, but the chemicals that make up these retardants are themselves toxic and may be triggering a health crisis of their own. Fire retardants are found in baby and adult clothing, mattresses, car seats, fabric blinds, and other products that can cause a host of illnesses, especially in children.

The chemicals added to these products are meant to make them harder to burn, but they are not only often ineffective but also toxic to health because they may leach from the product and become a part of everyday household dust. If that dust is touched or inhaled, it is now a bigger health threat unless the hands or affected areas are thoroughly washed right away. Newer organophosphorus compounds such as organophosphate ester flame retardants (OPFRs) have a low water solubility and bind to soil and dust.

These retardants are also hard to get rid of in the environment and have a poor level of biodegradation. They possess a higher vapor pressure and a shorter half-life than previous chemical compounds, but because they are more soluble and can persist in water, they are showing up in abundance in environmental tests. Because they are being used in such large amounts in many products, they are presenting global health problems to humans, especially children, and as of now, there are no regulations over the specific compounds involved. One of the most negative effects is on the brains of children. High levels of OPFRs have a negative impact on healthy brain development in children and on adult fertility.

Because the added chemicals make you sick and don't really meet flammability regulations, it's almost pointless to use them because they don't retard fire, but they do leach into your body. Ironically, firefighters are at the front line of those asking for a ban on such products because the toxic fumes they produce during a fire put first responders at a higher cancer risk.

 Because they [flame retardants] are being used in such large amounts in many products, they are presenting global health problems to humans, especially children....

One simple way to lessen their impact is to wash your hands. A June 8, 2020, report in *Environmental Science and Technology Letters* stated that hand washing could reduce exposure to the chemicals in flame retardants that are often transferred to the skin of the hands after touching products that use retardants, including cell phone cases, television sets, tablets, and, of course, clothing.

Food Additives

According to James E. Marti, our food contains over 1,000 intentional additives, many of which are carcinogenic, such as nitrite. In addition, there are over 12,000 unintentional additives in food production, processing, and packaging, including vinyl chloride and diethylstilbestrol, both suspected carcinogens. The main use of additives is to improve taste, texture, and increase shelf life.

Not all additives are added in later to the processing. Plant foods often contain certain molecules designed to protect them from insects and microorganisms, some of which are carcinogenic and mutagenic in humans. Black pepper contains piperine and sarole, herbal teas contain pyrolizidine, mushrooms contain hydrazines, and celery when bruised contains psoralen. The risks from food additives run from minor allergic reactions and irritations to cancer.

In 2019, the American Institute for Cancer Research warned that grilled foods increased the production and consumption of cancer-causing chemicals. Grilling meat at high temperatures releases potent, cancer-causing chemicals called heterocyclic amines (HCAs). When the meat (or fish or poultry) is grilled, these chemicals are released, as well as polycyclic aromatic hydrocarbon (PAH), which are formed when fat and juices are grilled directly over a heat source or heated surface and cause flames or smoke. In the smoke, the PAHs adhere to the meat, which is then consumed.

MSG

Monosodium glutamate (MSG) is one of the most notorious food additives because of its ability to enhance flavor and make us want more of the foods that contain it. This includes Chinese food, soups, processed meats, chips and snacks, and canned vegetables, and though the FDA has labeled MSG "generally safe," there are a host of associated symptoms people have reported from consuming it. It is derived from the amino acid glutamate, or glutamic acid, which is abundantly found in nature. In its chemical form, it resembles table salt and combines sodium and glutamic acid, which is made by fermenting starches. Though there is no real difference between this and the natural version, the glutamic acid in MSG is considered more absorbable because it is not bound inside a big protein molecule in the body. It also goes by about a dozen aliases, including glutamic acid, hydrolyzed protein, autolyzed protein, textured protein, autolyzed yeast extract, and maltodextrin.

Among the symptoms caused by MSG are headaches, chest pain, weakness, nausea, flushing, sweating, rapid heartbeat and palpitations, facial tightness, and numbness and tingling in the face and neck. These symptoms are considered anecdotal, as researchers have not found a definitive link between them and MSG, but some researchers suggest there are people who are sensitive to the product even in small amounts. Glutamic acid is a neurotransmitter in the brain and an excitatory one, which means it excessively stimulates nerve cells. MSG has been labeled an excitotoxin because of this, and some of the anxiety over consuming it comes from a 1969 study that found injecting large amounts of it into newborn mice caused them to become excitable.

Monosodium glutamate is a flavor enhancer often used in Asian food. A crystalline substance, it has no flavor of its own, but scientists believe it stimulates glutamate receptors in the tongue to make foods taste better. Many people are allergic to MSG, however.

Large doses of MSG can raise blood levels of glutamate, but dietary glutamate is not found to cross the blood–brain barrier in big enough amounts to cause problems. Still, there is plenty of firsthand experience from people who have had bad reactions to MSG, other small studies have linked it to an increase of asthma, and others still have linked it to obesity and weight gain.

Use it at your own risk.

Preservatives

Foods contain a variety of chemical preservatives designed to stop mold and bacterial growth, extend shelf life, and keep food fresher for longer. The preservatives are synthesized in a laboratory (some do occur naturally such as lemon juice, salt, and sugar) and then added to foods like processed meats, processed baked goods, sodas, wine, salad dressings, frozen foods, fruits (they keep them from spoiling), cereals, and more. People have reported a variety of symptoms such as allergic reactions, rashes, skin breakouts, headaches, breathing difficulties, and fatigue from food additives.

> People have reported a variety of symptoms such as allergic reactions, rashes, skin breakouts, headaches, breathing difficulties, and fatigue from food additives.

Preservatives are broken down into three categories:

1. Antioxidants that slow down oxidation of fats and prevent foods from going rancid.

2. Antimicrobials that stop mold, bacteria, and yeast from growing and spoiling foods.

3. Preservatives that inhibit rot in produce.

Names of preservatives show up on labels as:

▲ Benzoates: Found in salad dressings, pickles, and milk products, they kill bacteria and other microorganisms that grow in acidic conditions. Sodium benzoate, found in soft drinks, is linked to DNA damage.

▲ Sorbates: Potassium sorbate, calcium sorbate, and sodium sorbate. Sorbic acids act as antimicrobial agents and are used in cookies, sodas, frozen pizzas, and other frozen foods.

▲ Butylated hydroxyanisole (BHA): A waxy solid that preserves butter, lard, processed meats, and the like. Linked to cancer and tumor growth.

▲ Butylated hydroxytoluene (BHT): The powdered form of BHA, which has been banned as a carcinogen in many countries. It prevents oxidation of fats in food, which then turns food rancid and is found in cereals, snacks, processed foods, and baked goods.

▲ Sulfites: Sodium sulfite, sodium bisulfite, sodium metabisulfite, potassium bisulfite, and potassium meta-bisulfite. Antioxidants that inhibit oxidation.

▲ Calcium sulfate: This regulates acidity in foods, allowing them to last longer under certain acidic conditions. It also is used as a flour stabilizer in processed foods and breads.

▲ Chelating agents and binders: Disodium ethylenedia-minetetraacetic acid, which binds metal ions to prevent oxidation, and polyphosphates, which are found in peeled fruits, peeled veggies, and dips.

▲ Methylcyclopropene: This is a gas mainly used to preserve fruits and vegetables in crates by inhibiting the production of ethylene gas, a natural hormone that ripens food faster. This preservative can keep a banana fresh for up to a month and an apple up to a year.

▲ Nitrates and nitrites: They generate nitrosamines when mixed with stomach acid during digestion and are considered carcinogenic. They are mainly used to preserve meats and destroy bacteria and microbes that cause food spoilage. Even if the package says no nitrates/nitrites, keep in mind the manufacturer can use them by claiming they're cured with natural nitrates/nitrites. The compounds are chemically identical, though, and have the same effects on health.

▲ Dimethyl decarbonate: This is another chemical additive that inhibits the growth of enzymes which, in turn, prohibits the growth of microorganisms. These are mainly found in sodas and beverages like ice teas and sports drinks and are, along with sulfur dioxide, one of the most common preservatives found in wine.

Sulfites in Wine

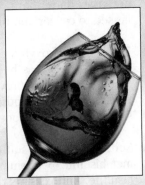

In wine making, sulfites help maintain the flavor and freshness of wine. But they are associated with side effects, including the notorious "wine head- ache," and can also cause hives, swelling, and stomach pain. Their anti- microbial properties are what help wine stay fresh, and although they are found in other foods such as black tea, juices, jams, jellies, sausages, eggs, peanuts, and dried and pickled foods to keep them from spoiling, their presence in wine helps to keep the liquid from browning, which alters not just color but flavor.

Potassium metabisulfite in wine comes from its use as a sanitizer for wine barrels and wine-making equipment.

According to the FDA, about 1 percent of people in the United States are sensitive to sulfites in wine. For people with asthma, sul- fites can irritate the respiratory tract. Obviously, the more wine you consume, the greater the potential of getting a reaction, such as a headache. There are wines on the market that are sulfite-free, but bear in mind, drinking a whole bottle might still cause a headache!

In small amounts, these chemicals may not be super harm- ful, although some have been tagged as carcinogenic or potentially carcinogenic, but because we humans tend to overeat junk and pro- cessed foods and beverages, the amount of these chemicals in our bodies adds up. Eating organic or looking for preservative-free choices, which must be consumed sooner to avoid spoiling, are good options to avoid toxic buildup.

Artificial Sweeteners

In our quest to avoid eating more sugar, which itself causes health problems, we have adopted a host of artificial sweeteners that are nothing but chemicals and more damaging to the body. The most widely known and used are aspartame, acesulfame potassium (Ace K), sucralose, trehalose, and sorbitol.

The use of these products allows manufacturers to call their foods "diet" foods, which is incredibly deceiving to the consumer concerned with overall health. Sucralose can show up in foods that are not necessarily diet friendly. Sucralose and Ace K are found in things like diet soda and can cause weight gain because people assume they are "healthier" than regular sugar and consume more of the products. But these chemicals lead to increased inflammation in fat cells, which allows excess glucose to be stored in the body as fat, and they lead to the creation of more fat cells.

Trehalose is a sugarlike flavor enhancer that adds a little sweetness to foods, extends shelf life, and improves texture. It is found naturally in mushrooms, yeast, and shellfish, but in processed foods, it is used in higher levels and can cause inflammation of the colon, increased diarrhea, and might even be deadly because of its ability to encourage bacteria growth.

Sorbitol is sugar alcohol that has half the calories of regular sugar and is not as sweet. It is used in sugar-free versions of baked goods, candy, and gum. But because it introduces excess water into the colon, it acts as a laxative and causes gas, bloating, and diarrhea.

Aspartame is a nonsaccharide sweetener that is 200 times sweeter than sucralose and is commonly used in foods and beverages that go by the name Equal®, Canderel®, and NutraSweet®. Most diners and restaurants offer these products with coffee and tea, and they are also found in sodas and sweet drinks. Aspartame is an aspartic acid and phenylamine dipeptide, a long name that carries long-term repercussions when consumed. Aspartic acid and phenylamine are naturally occurring amino acids, but when your body processes aspartame, it breaks it down into methanol. As of 2014, aspartame was the largest source of methanol in the American diet, and it is toxic in large quantities.

 Aspartic acid and phenylamine are naturally occurring amino acids, but when your body processes aspartame, it breaks it down into methanol.

When heated, it creates free methanol, and when combined with regular methanol, it is especially toxic because it breaks down into formaldehyde, which we know is a neurotoxin and a carcinogen. There is a lot of controversy over consuming aspartame, and because you do get methanol from other parts of your diet, such as fruits and vegetables, it has not been considered a high priority for research into potential harm.

Aspartame has the approval of the FDA, the UN Food and Agriculture Organization, the American Heart Association, and the WHO. A 2013 European study of over 600 datasets concluded it was safe. But the FDA has banned many artificial sweeteners, such as saccharin and cyclamate, and there have been studies by the Harvard School of Public Health and Center for Science in the Public Interest showing aspartame was not as safe as previously thought. Some of the issues with products like these come from lack of studies and pressures from the industry to keep the product on the market shelves, so the consumer may not be getting the whole story.

Activists and health groups insist aspartame is responsible for cancer, headaches, depression, seizures, attention deficit hyperactivity disorder (ADHD), birth defects, weight gain, lupus, and multiple sclerosis (MS), among other health ailments, and it is up to the consumer to decide if a product so surrounded in controversy on both sides is worth the risk, especially when there are natural sweeteners like stevia, molasses, honey, and agave nectar.

Phosphates

This form of phosphorus is a mineral that supports bone health. Phosphorus-containing additives such as phosphoric acid and disodium phosphate are found in everything from sodas to baked goods, dairy products, fast foods, and processed foods. High levels can cause damage to the kidney function because of the extra work the kidneys must do to excrete the phosphorus. It also binds to calcium and then pulls it from the bones, leaving bones weak and brittle. In a 2013 study in the United Kingdom with more than 700,000 participants, those with normal kidney function and higher phosphorus levels had a 36 percent higher chance of stroke or heart attack over those with normal phosphorus levels.

Carrageenan

This food stabilizer comes from seaweed and is used in things like salad dressings to keep the ingredients from separating. It also gives yogurts and frozen desserts a nice, creamy texture. The problem is, it has a structure that is foreign to human cells and causes inflammation in the body, often in the gastrointestinal tract, and can cause irritable bowel syndrome. It also has been linked to other inflammatory diseases like cancer, diabetes, and arthritis.

Though many watchdog groups like Consumer Reports have called for a ban on carrageenan, the USDA has instead chosen to allow it to be used even in organic products, although many organic food companies have chosen to stop using it on their own. It still shows up in a wide variety of products, so check those labels.

Food Colorings and Dyes

Americans eat five times more food dye now than they did in 1955, according to "Colors to Die For: The Dangerous Impact of Food Coloring," published by Special-Education-Degree.net. The chemicals used to make food more attractive contain many hidden dangers, especially to children, who consume them most in the form of frostings, candies, and snack foods. We like our food to be pretty, and a perfect example is farm-raised salmon, which is naturally an ugly, gray color. Who would eat it? So, pink coloring is added to make it nice looking and to match how wild salmon looks naturally because of the abundance of astaxanthin, a red-orange compound found in krill and shrimp.

In 2009, the United Kingdom asked food companies to stop using artificial dyes after finding links to cancer and hyperactivity, and today, Kellogg's and Kraft Foods no longer use the chemicals there. In the United States, however, the FDA states there are no compelling studies to show links to child behavioral issues like ADHD despite several studies to the contrary. Several studies published in *Science* linked these colorants to lower test scores, lower memory recall, and hyperactivity, which is a shame because there are many natural food colorants on the market that can be used instead.

Here is a quick rundown of the food-coloring dyes:

▲ Blue #1 (Brilliant Blue): Linked to kidney tumors in mice, allergic reactions in individuals with asthma, and found in baked goods, cereals, packaged soups, canned peas, popsicles, icings, beverages, and birthday candles.

▲ Blue #2 (Indigo/Carmine): Linked to brain gliomas in male rats. Found in beverages, candles, candy, ice cream, snacks, and dog food.

If you're eating farm-raised salmon, that lovely pink coloring is the result of food dyes, not the natural krill diet that wild salmon consume.

▲ Red #2 (Citrus): Linked to bladder tumors in mice and labeled a possible carcinogen by the WHO's International Agency for Research on Cancer (IARC). Found in Florida orange skins.

▲ Red #40 (Allura Red): Linked to immune system tumors in mice, allergic reactions, and hyperactivity in children. Found in cereals, beverages, candles, sports drinks, condiments, and some cosmetics.

▲ Red #3 (Erythrosine): Linked to thyroid cancer in animals, hyperactivity in children, and partially banned in 1990 by the FDA. Found in baked foods, maraschino cherries, popsicles, cake-decorating gels, candles, and some sausages.

▲ Green #3 (Fast Green): Banned in the European Union (EU) and many other countries for significant increases in bladder tumors in male rats. Found in ice creams, beverages, and candles. Also used in some cosmetics.

▲ Yellow #5 (Tartrazine): Linked to allergic reactions and mild to severe hyperactivity. Found in baked goods, cereals, chips, popcorn, beverages, and candles. Also found in cheese to make it yellow.

▲ Yellow #6 (Sunset Yellow): Linked to adrenal tumors in animals, hyperactivity in children. Found in cereals, beverages, sauces, preserved fruits, and candles.

 Despite FDA approval, the above are still linked to many health hazards, leading one to wonder how broad the FDA considers the word *generally*....

The FDA gives what is called a "generally regarded as safe" (GRAS) rating to good dyes, which is why you do not see Red #4 anymore on shelves, as it has been banned. Despite FDA approval, the above are still linked to many health hazards, leading one to wonder how broad the FDA considers the word *generally* when determining safety levels. In a study called "Toxicology of Food Dyes," published in the July-September issue of the *Journal of Occupational Environmental Health*, the conclusion was that many dyes raised varying degrees of concerns. Red #3 and other dyes were carcinogenic, and Red #40, Yellow #5, and Yellow #6 were contaminated with benzidine, a carcinogen. Four dyes—Blue #1, Red #40, Yellow #5, and Yellow #6—were definitively linked to hyperactivity.

Reading labels helps, as you can find products that use natural colorings such as saffron powder, red cabbage, pumpkin, carrots, paprika, beets, and spinach. The color may not be as brilliant and concentrated as chemically colored foods, but it's a lot healthier.

Meat, Fish, Dairy, and Additives

Factory and intensive farming for meat and dairy to keep up with the public's growing desire for everything from steak to salmon to eggs and cream has resulted in numerous methods for increasing output, size, and efficiency of operation. In 2018, approximately 80 percent of all U.S. cows were injected with growth hormones to increase their size, their lean meat ratio, and their feed-use efficacy. U.S. cattle, sheep, and other meat animals are injected or implanted with hormones unless the package is marked USDA Organic Certified.

Since the 1950s, the FDA has permitted the use of steroid hormones in the form of pellets or implants that are placed under the skin of the animal's ear. They dissolve slowly and do not need to be removed. The ears of a treated animal are thrown away and not allowed to be consumed (right there is a red flag!) and have a "zero-day withdrawal," which means the meat of a treated animal is safe to eat anytime after the implant or treatment. The hormones cause the animal to gain weight more quickly or, in the case of dairy cows, produce more milk.

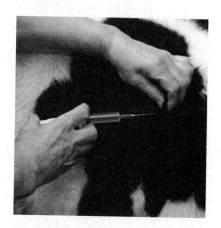

The three most commonly used hormones are natural: estradiol, progesterone, and testosterone. These are growth and sex hormones, and even though they are found in the human body, adding them to the meat we eat means we are getting more of these hormones than we would naturally. Chemical hormones such as trenbolone, zeranol, and melengestrol acetate are fed to cattle, too. Hormones are broken up into two categories: steroid and protein. Steroid hormones go directly into your bloodstream and to your cells. Protein hormones are broken up into

Farmers in the United States inject dairy cows with a variety of hormones to improve meat quality. Recently, researchers at the University of Wisconsin at Madison discovered that serotonin injections increase calcium content in milk.

smaller molecules and digested in the stomach, where they lose much of their potency.

As usual, the FDA claims these hormones are not found in meats at unsafe levels, but there is plenty of research to show otherwise. Studies are showing that adding more hormones to the body is not good for pregnant women and is causing females to enter puberty at earlier ages than ever before. Despite the FDA regulating hormones in meats using the maximum residue level (MRL), there are higher MRLs for certain types of meat than others. Organ meat has higher levels than muscle meat.

Dairy cows are fed hormones to increase milk production. Since 1993, the FDA has allowed recombinant bovine growth hormone (rbGH) to be injected under the skin of dairy cows. The Cornell University Sprecher Institute for Comparative Cancer Research claims the amount in milk is not linked to any bad health effects, but higher levels have been found in women with breast cancer, according to a study from the Program on Breast Cancer and Environmental Risk Factors in New York state. Steroid hormones in general are a cause of public concern over possible links to breast cancer and early onset puberty in girls. There are few studies that show direct links between added hormones and human illness or disease, but advocates against their use claim the presence of such hormones in meat and dairy can cause breast growth in boys, cancer, and disruptions in development.

In a 2010 study published in the *Journal of Clinical Investigation*, Sara Divall of Johns Hopkins University found that giving mice inulin-like growth factor (IGF-1), a hormone that promotes growth, caused the mice to enter puberty earlier than usual. The European Union has banned the use of rbGH and other growth hormones in their dairy and meat.

There are foods labeled hormone-free, and buying from a known and trusted local farm or ranch is a great way to avoid these additives. Be careful with "organic" meat and dairy because the label does not prevent manufacturers from adding hormones.

The same concern goes for added antibiotics into meat such as roxarsone, which is commonly used in factory farms and contains known carcinogens, including arsenic. The USDA claims that eating two ounces of chicken per day exposes consumers to three to five grams of inorganic arsenic, which is its most toxic form. The presence of added antibiotics causes a real problem in a populace already dealing with the overuse of antibiotics, which makes people far more susceptible to superbugs that are antibiotic resistant.

 There are foods labeled hormone-free, and buying from a known and trusted local farm or ranch is a great way to avoid these additives.

It's not just what is injected into these animals, though. It's also what they carry, such as salmonella, *E. coli*, and other bacteria. It's also what they *eat*, as in the kind of feed they are fed, which often includes pesticides, toxins, and bits and pieces of other animals.

Is Fake Meat Better?

Fake meat has become all the rage today, with product lines such as Beyond Meat getting into production of fake beef, pork, and chicken products made in China. These products often claim to be a healthier alternative to real meat because they are plant based, but they are not telling the truth. These foods are ultraprocessed and filled with ingredients people might not be aware of. Beyond Burger patties contain twenty-two different ingredients to give them the taste and texture of "almost meat."

Foods that come from China are not known for their quality or safety, so that should be the first red flag. If these foods are chemically contaminated in the production process, consumers on the other side of the world would never know. But ultraprocessed foods can be harmful to human health. According to the NOVA Food Classification System created by the Center for Epidemiological Research in Health and Nutrition, these products are "industrial formulations made entirely or mostly from substances extracted from foods (oils, fats, sugar, starch, and proteins), derived from food constituents (hydrogenated fats and modified starch), or synthesized in laboratories from food substrates or other organic sources (flavor enhancers, colors, and several food additives used to make the product hyper-palatable)."

Doesn't sound very healthy, does it? Another big fake meat company, Impossible Foods, has over a dozen patents with hundreds more pending for its foods, which is enough to raise the hair on the back of your arms. Natural foods cannot be patented. Fake formulations can. Many of these patents are for methods and content that change the taste, smell, or texture of a product. These "plant-based meats" are not natural by any stretch of the imagina-

Meat substitutes such as Beyond Meat combine plants with other ingredients to simulate the texture and taste of real meats.

tion. Nor do they specifically protect the environment, as many companies claim, by being better than meat production because they are comparing their carbon footprint to factory farms, not open-range and free-roaming, small farms. By comparing themselves with the worst of the worst (factory farming), they come out looking great for health and the environment.

The Impossible Burger contains genetically modified soy and eleven times higher levels of glyphosate than Beyond Burgers, which also contains the toxin.

You're much better off eating real meat that is grass-fed and from a small local farm or ranch. They are better not only for your health but for the environment as well.

Farmed Fish

It isn't just cows, pigs, and sheep. Farmed fish contain high levels of mercury, a potent neurotoxin, and many are contaminated with PCBs. These fish are fed food filled with persistent organic pollutants (POPs). All of this ends up in the bodies of humans. The fish oils made from these fish are also highly contaminated, and many become the fish oil supplements sold in stores and online to consumers who think they are doing something good for their health.

Because farmed fish have more fat than wild caught fish, it has been discovered that the extra fat is chock full of DDT and other toxins. DDT was banned back in 1972 but remains in the environment enough to still be contaminating everything today, such as fish, soil, and plants. One of the biggest problems these toxins pose is the fact that they have half-lives ranging from a few months to a few years and cause havoc to the body during that time frame. Eating farmed fish once a week is enough to cause a toxic load that will have long-term repercussions.

Wild caught fish like salmon and sardines have less mercury and toxins and are high in Omega-3 fatty acids, which are extremely beneficial to human health. But sadly, as we dump more toxins and pharmaceuticals into the environment, more will make

Mercury and Dental Fillings

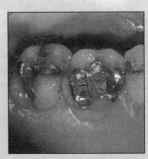

The debate over the dangers of mercury fillings in teeth has raged on for decades. Mercury, which we know is a potently dangerous neurotoxin heavy metal, has been linked with fatigue, brain fog, confusion, neurological issues, and even dementia and Alzheimer's. In dental terms, mercury fillings are called amalgams, or "silver fillings," and are made up of 55 percent mercury. Each dental filling can release about one microgram of mercury every day, which is absorbed into the body. If you have a mouthful of fillings, do the math. That's a lot of mercury exposure on a daily basis.

The amount of mercury in the body shows up in urine, and according to the CDC, mercury has a half-life of two months, which is the measure of time it takes for 50 percent of the substance to be cleared out of the body. It is hard to get rid of, and in pregnant women, it can make its way to the fetus, with the fetal brain containing approximately 40 percent more mercury than the mother's brain. The mercury that passes into the fetal brain comes from the mother's mercury fillings and can cause neurological damage to the fetus. Even worse, our capacity to get the mercury out of our brain is limited so that once the limit is reached, the rest of the mercury cannot get out. As the toxic load on the brain increases, the ability of the body to detox it gets even harder.

In September 2020, decades after the FDA was pressured to pay attention to the dangers of amalgam fillings, it released its safety communication on amalgam, stating that the mercury vapor released by these fillings adversely affects pregnant women, nursing women, newborn babies, people with impaired kidney function, people with neurological diseases such as Parkinson's and Alzheimer's, children, and other susceptible individuals and should be avoided by these groups. This shocking admission came about from a lawsuit by Consumers for Dental Choice, an organization that sued the FDA, and with the help of doctors, researchers, and activists, it put the heat on the FDA to come out and make this statement and to push for an eventual overall ban of the use of mercury. The FDA was also pushed to tell dentists to not use the deceptive term *silver fillings* with patients, as this implies the fillings are not made of mercury.

their way back to us. The juvenile salmon caught in Washington state's Puget Sound are said to be filled with common drugs, including Flonase, Aleve, Tylenol™, Paxil, Zoloft, OxyContin, Darvon, Cipro, and more. This means that no matter how good and healthy the water you drink is, when consuming fish that live in areas with wastewater and runoff, you could be taking a lot of medications you don't even know you're taking.

Mercury remains the biggest toxic problem, with tilefish, swordfish, shark, and king mackerel showing the highest levels, followed by fresh and frozen tuna, grouper, trout, red snapper, and lobster from the northern United States.

Organic or Not?

Customers looking for organic milk might be surprised to learn that in 2019, the U.S. dairy sector was being investigated for making fraudulent organic claims. The USDA was given evidence of cows that were being passed off as organic at the Erath County Dairy Sales in Texas. Apparently, the Erath County Dairy Sales was found removing cows' ear tags and replacing them with "back tags," which falsely indicated a cow is organic. By doing so, they can increase the meat sale value by three to four times.

These ear tags not only fool the consumer into thinking they are eating organic, but they also prevent the traceability system of an animal's movements from the farm to auction to the slaughterhouse to consumer. This means there is no way to know where the meat or dairy came from. Not knowing this also means it would be impossible to trace back animals that may have been exposed to deadly viruses or bacteria.

The organic meat and dairy industry is vulnerable to this, especially since about 60 to 70 percent of agricultural products imported into the country under the USDA-certified brand are potentially fraudulent, and without shortening the supply chain and demanding organic certification from each party along the supply chain, it will no doubt continue. The only way around this is to buy local and support local and family farms because the organic meat and dairy at the local grocery chain could be anything but.

Wrap It Up

Buying foods wrapped in plastic, and cooking and storing with plastic wrap, presents another hazard. This includes wax paper and liners. The chemicals that make these papers not stick to food and pliable are also toxic, including endocrine disruptors that cause hormonal issues, raise blood pressure, and increase risk of cardiovascular disease. Plastic storage containers are cheap, but the lids often contain bisphenol A. (BPA), so do your health a favor and cook and store with glass.

This applies to condiments, too, like peanut butter, catsup, nut butters, mustard, and salad dressings. Avoid plastic, and look for those that come in glass jars or containers. Once you use up the product, thoroughly wash the jars and lids and use them for food storage. Look for products in soda-lime or

Some unscrupulous companies have removed ear tags from cows, making it impossible to trace their origins and easily conceal if they have been exposed to viruses and bacteria.

borosilicate glass, which are more able to withstand heat and cold. If you choose to use plastic, you should always seek out those that are BPA-free.

Furniture

Just as the bedding you sleep in each night contains toxins, so does the furniture in your home and office. Depending on the materials, there are a host of toxic chemicals involved in the making, production, and finishing of furniture. Upholstered furniture contains polyurethane foam, which contains flame retardants. Those items made before the mid-2000s may contain polybrominated diphenyl ethers (PBDEs), which cause thyroid problems, neurological impairment, and fertility. Anything upholstered that has been Scotchgarded at one point adds even more chemicals to the mix. You can check the labels on the items to see if they are flame-retardant-free, but other chemicals may not be listed.

Wood furniture, especially the cheaper kind found in big-box stores and discount furniture stores, can bring a number of toxins into the home or office. One of the biggest is formaldehyde, which is used to preserve the wood and is also found in particle board, plywood, and fiberboard. You will also find formaldehyde in the coatings, lacquers, and paints used on wood products, and few people realize that this chemical turns to a gas at room temperatures and then is inhaled. It is always best to keep windows open, especially with newly purchased wood furniture, to help dissipate high levels of the toxin in the air. It also helps to buy the higher quality, more expensive wood products because they contain fewer toxins. Most furniture will have a label that identifies the formaldehyde risk, so look for ultralow-emitting formaldehyde (ULEF) or no added formaldehyde (NAF).

Wood cabinets, bed frames, bookshelves, doors, desks, and other products are one thing, but even leather furniture is chock full of formaldehyde and other leather-tanning treatments.

Hazards lurk in the kitchen, too. Granite countertops are hugely popular but can be a big source of radon in the home.

Hair Dyes and Salon Treatments

Millions of men and women use dyes on their hair. Men even dye their beards. This quest to slay the grey has resulted in numerous studies linking hair dye with cancer, not to mention scalp and skin irritation. There are different kinds of hair dye:

Temporary: These cover grey by coating hair strands but only last for a few washings.

Semipermanent: These penetrate the hair shaft and last for up to ten washings.

Permanent: These cause complete chemical change of the hair and last for longer periods of time. Many contain coal tar and all contain colorless toxins like aromatic amines and phenols. They also contain hydrogen peroxide, which is what causes the chemical reactions in the dye to make it work.

A rule of thumb is that the darker the hair dye, the more chemicals it likely has in it. These chemicals can be inhaled or absorbed through the skin.

The darker the dye, the more chemicals it contains. Small amounts are absorbed into the skin and inhaled, whether done at home or in a salon. According to the American Cancer Society, studies show a link between hair dye and cancer, specifically breast cancer, bladder cancer, leukemia, and non-Hodgkin lymphoma. They also show that people who work in salons and are exposed to these toxins in greater amounts have more risk of bladder cancer, but this risk did not appear to be the case in those who were getting their hair dyed.

Women who began using hair dye before 1980 and those who consistently use black and dark brown colors had a slightly elevated risk for leukemia and non-Hodgkin lymphoma. Links have been made to breast cancer, but most of the studies show that risk to be small for those who use dye occasionally. The International Agency for Research on Cancer found that based on the data regarding bladder cancer, there is evidence to show that there is higher workplace exposure, such as with hairdressers and barbers, to these possible human carcinogens, but the organization has not gone so far as to label hair dyes as reasonable human carcinogens.

The National Toxicology Program (NTP), which consists of several government agencies such as the FDA and the National Institutes of Health (NIH), have not classified hair dye exposure as linked to cancer, but it has classified some of the chemicals in hair dye as "reasonably anticipated to be human carcinogens."

This kind of vague-speak is confusing for concerned consumers, who may benefit from using the vegetable-based dyes that may not be as long-lasting and dramatic but have no damaging chemicals. Other studies show that hair treatments can harm the nervous system because of the presence of toluene, a clear, colorless liquid with a strong smell that acts as a solvent to dissolve other products, and increase the risk of peripheral neuropathy, including one study in the February 12, 2013, issue of *Neurology* that looked at men who dyed their beards and found chronic lead toxicity from long-term exposure to dye toxins that were orally absorbed from the beard.

The American Cancer Society claims that hair dyes and treatments absorbed into the skin can cause allergic reactions and skin and eye irritations, including rashes and welts, especially around the hairline, and swollen eyelids. Some women attest that the dye made their hair fall out (it's always good to check public forums for stories from actual users).

 Perm chemicals and other hair-straightening or hair-waving treatments include powerfully toxic compounds that get into the bloodstream with long-term use.

It isn't just dye and coloring treatments that are chock full of chemicals. So are products used in wigmaking, wig glue, hairpiece bonding, hair-extension glue, lace wig glue, and one of the most notoriously chemical-laden treatments of them all—the perm. Perm chemicals and other hair-straightening or hair-waving treatments include powerfully toxic compounds that get into the bloodstream with long-term use. One of these chemicals is toluidine, which is a carcinogen. Again, those who work in wig shops, barber shops, and salons risk the most exposure, while the customer encounters the chemicals for a shorter duration.

Other toxic chemicals found in salons, including nail salons, are:

▲ Dibutyl phthalate: found in nail polish; causes dizziness, nausea, eye and skin irritation

▲ Formaldehyde: found in nail polish, hardener, keratin hair treatments; causes respiratory issues, headaches, wheezing, skin and eye irritation, throat irritation

▲ Methyl methacrylate (MMA): found in artificial nails; causes breathing problems, headaches, eye irritation, confusion

▲ Styrene: found in hair-extension glue, lace wig glue; causes vision problems, fatigue

▲ Trichloroethylene: found in hair-extension glue, lace wig glue; causes confusion, headaches, nausea, eye and skin irritation

▲ 1,4 dioxane: found in hair-extension glue, lace wig glue; causes nose and eye irritation

▲ Cyclomethicone: found in flat-iron spray, heat-protection spray; exposure to heat causes increase in formaldehyde and breathing problems, wheezing, rashes, eye and throat irritation

▲ Acetone: found in nail polish remover, hairspray; causes eye, throat, and skin irritation and dizziness

▲ Utyl acetate: found in nail polish, polish remover, wig and hairpiece bonding; causes dizziness, headaches, eye and skin irritation

▲ Methacrylic acid: found in eyelash glue, nail primer; causes skin rashes and burns, eye and throat irritation

▲ Glyceryl thioglycolate: found in permanent-wave solutions and acid waves; causes skin irritation and dermatitis

▲ P-phenylenediamine: found in hair dyes, henna tattoos; causes skin irritation and rashes

▲ Ammonium persulfate: found in hair bleach; causes asthma, shortness of breath, skin and eye irritations

The above list should scare anyone from ever going to a salon again, but keep in mind not all salons use the same products, and many do have lines of natural and/or organic products. Also keep in mind you'll be spending a lot less time there than the people who work there, so your exposure will be minimal by comparison. At the very least, always look for a salon that is well ventilated, and ask about the ingredients in the products they are using. When treating hair at home, there are natural and temporary hair dyes and less chemical-laden and less harsh treatments that won't sacrifice health for beauty. Even nail polishes offer fewer toxic alternatives, or you can choose to buff to a shine and go natural more often and give your skin, eyes, nose, throat, and lungs a break.

Microwaving popcorn seems like a harmless enough way to cook a tasty snack, but the bags contain PFCs, which break down into PFOA, which can cause Alzheimer's.

Microwave Popcorn

Everyone knows the dangers of eating processed snack foods for their poor nutritional content, added salts, colorings, and flavorings, and way too much sugar. Popcorn has long been thought to be a healthy snacking alternative, but in our busy, hectic world, we have made this product dangerous to health, especially children's health. Buying the quick and easy microwave popcorn in a box seems like a great idea, but the bags used are lined with a thin coating to prevent the leaking of oil that contains PFCs, which break down into perfluorooctanoic acid (PFOA), which is a toxic chemical linked to cancer of the testicles and kidneys, thyroid issues, and ADHD. PFOAs take quite a long time to get out of the body once they are introduced to the bloodstream.

In the early 2000s, reports were in the news about popcorn factory workers experiencing serious health problems, mainly a lung disease called bronchiolitis obliterans, which became known as "popcorn lung." Research revealed the culprit was the chemical diacetyl, used by most major brands to give popcorn its buttery flavor and aroma. Diacetyl is found in many foods in small quantities, but when heated in a microwave, the chemical vaporizes and becomes toxic.

Most major brands removed diacetyl after the stories broke, especially after researchers from the University of Minnesota found in a 2012 study that diacetyl passed the blood–brain barrier and caused brain proteins to "misfold" into beta amyloid, which is one of the primary pathologies for Alzheimer's disease. It gives the consumer pause, though, to think about all the other chemicals in our foods that we are not being alerted to because nobody reported problems with them. It is tragic that the people who work in the manufacturing plants and production warehouses are never told either until it's too late and their health has been damaged beyond repair.

 Popcorn is eaten by the bowl and is a favorite children's snack, so it is consumed in large quantities without much thought....

These products also tend to use basic table salt, flavorings with chemicals, and the worst kind of oils—hydrogenated. They also contain trans fats, which are linked to cancer, heart disease, and obesity. Popcorn is eaten by the bowl and is a favorite children's snack, so it is consumed in large quantities without much thought, but it might behoove adults to make it themselves by putting dry kernels in a paper bag, taping the bag, and popping it in the microwave, then sprinkling with a fine olive oil mist and some quality salt. This avoids the chemicals from the bag and the damaging oils. Even better, learn how to pop it in a giant pot, the old-fashioned way.

Paint

A fresh coat of paint can brighten up any room, but it can also cause headaches, dizziness, nausea, respiratory issues, and even kidney, liver, and nervous system damage if inhaled long enough, a problem experienced by professional painters and paint spray booth workers. VOCs are mainly to blame because they evaporate at room temperature and are found in paint pigments, which also contain alkyd oil. Paints also include solvents and formaldehyde, which are dangerous when inhaled.

Paint toxins include:

▲ Xylene

▲ Ethyl acetate

▲ Methylene chloride

▲ Formaldehyde

▲ Glycol

▲ Benzene

▲ Phenol

▲ Xylene

▲ Quaternary ammonium compounds

Inhaling paint fumes can also cause trouble for those with allergies, chronic obstructive pulmonary disease (COPD), asthma, and chemical intolerances, and paint manufacturers don't always disclose the ingredients or additives on the paint cans. Depending on the paint's toxic substances and the length of "off-gassing" processes, the chemicals can lead to short- and long-term health issues, such as nose and eye irritation and nerve damage. The Poison Control center admits that paint, when used properly, can irritate eyes, nose, and throat but usually won't cause much more damage, although wearing a mask while painting is a good idea. They do state that when you breathe solvent paint fumes for too long, you can experience headaches, nausea, and dizziness. However, paint fumes and the poisons found in inhalants can cause brain cells to die and, thus, the brain shrinks. A 2014 study from the Harvard Center for Population and Development Studies found that workers exposed to paint, glue, or degreaser fumes experienced issues such as memory loss during retirement.

There are paint companies that make low-VOC paints and neutral and earth tones with less pigment, which emit fewer toxins, but these are harder to find. Newly painted rooms should be avoided for a few days at least, and there should always be good ventilation to keep from inhaling too much of the VOCs they give off.

Pet Treatments

Parasiticides are used to kill fleas and ticks on our pets, but they are toxic not only to your pets but to the environment as well. Many pet health strategies include yearly use of parasiticides, such as flea baths and dips, but the blanket use of chemicals is not in the best interest of your pet. Less may be more when it comes to pet protection, but heavy advertising and promotional campaigns in the media and via your local veterinarian push the use of such toxic products, and perhaps the overuse, and few consumers know the dangers.

Parasiticides can have side effects on our pets. A paper in the January 2020 issue of *Veterinary Journal* by Christopher Little of Barton Veterinary Hospital in Canterbury, England, and his colleague Alistair Boxall of the Department of Environment and Geography at the University of York looked at the environmental pollution from pet parasiticides and found that the "blanket prophylactic use" of these products often leads to overuse and more harm to the environment, when a more strategic approach with specific parasiticides might be better.

You can control the amount of toxins used on your pet to control fleas, ticks, or ringworm by using only the treatment needed for your specific needs.

The lack of flexibility in using such broad-spectrum products makes it difficult to pinpoint which might also be making a pet sick from adverse effects. There is a call among veterinarians and pet health advocates to stop using such all-purpose chemical cocktails and instead focus on each pet's individual needs (maybe only fleas, maybe only ringworm, etc.), which will lead to limits on their environmental impact. It also makes more sense to expose Fido or Fifi to fewer chemicals to get the job done because these insecticides enter the animal's bloodstream. The pet's liver is going to have to process the toxins, so the fewer the better. Depending on the pet's age, weight, and other factors, it might be worth looking into natural alternatives or at least asking your vet for a more specific approach. Even going back and forth between a natural treatment and a chemical treatment for fleas or ticks can reduce the toxic overload to the animal and to the environment.

Flea collars are outdated products with all the new treatments available, but people who continue to use them should know they contain two carcinogenic ingredients: tetrachlorvinphos (TCVP) and propoxur. Both can cause brain damage and cancer, and these ingredients can also affect the pet's owners. Since children play with pets more often, they, too, are exposed to such dangerous toxins, which can delay motor development and cause attention deficit disorder (ADD) and hyperactivity.

Salt, Please?

Maybe not so fast because a 2019 study done by Greenpeace in conjunction with *National Geographic* found that there are microplastics in salt. Their study looked at thirty-nine different brands from sixteen countries all over the world and found that thirty-six of the thirty-nine brands contained microplastics. It did not matter if the salt came from the sea, a lake, or was rock salt, it was still tainted. This included most major salt brands, and if you consider how often you consume salt, you get a real solid idea of the quantity of microplastics you're ingesting.

Himalayan salt appears to be the safest and healthiest, as do some brands of sea salt, but you, the consumer, must research these products to see if they contain something you really don't want on top of your steak or salad. One of the worst offenders? Morton Table Salt, which can be found on almost every table in America.

One of the safer salts you can use is Himalayan salt (pictured), which contains little or no additives. Kosher salts are also a good bet.

Sunscreen

Sunscreen is supposed to protect us from the effects of the sun's rays and their potential to cause skin cancer. Yet the product itself has come under fire for being a risk to health. Four main ingredients that pose harm have been identified in a 2019 FDA study: avobenzone, oxybenzone, octocrylene, and ecamsule. These chemicals were found in concentrations that exceeded the FDA maximum threshold after only four uses in a day. At this point, these chemicals become potentially a bigger problem than sun exposure.

These active ingredients also serve to enhance the ability of other chemicals to penetrate the skin, including pesticides, herbicides, and insect repellents, all of which are known endocrine disrupters. Oxybenzone is highly absorbable and has both neurotoxic and endocrine-disrupting effects, yet the FDA has not discouraged people from using sunscreen made with it—nor, for that matter, have most dermatologists.

A January 2020 study by the FDA looked at these ingredients, as well as new ingredients found in sunscreens such as homosalate, octisalate, octinoxate, and other toxins in aerosol sprays and pump spray products. These ingredients are linked to hormone disruption, cell damage, endocrine disruptions, and cancer. Even in this newer study, the biggest offender was oxybenzone, found to be in concentrations of approximately 400 to 500 times above the safety threshold after a couple of days of use.

One of the most frightening aspects of toxins in sunscreen is that they are directly absorbed into the skin and into the bloodstream. They bypass the digestive tract that would otherwise eliminate the toxins. A little vitamin D from the sun is healthy and beneficial, but if you must use a sunscreen, read the ingredients, and look for something with more natural products such as mineral-based sunscreens. At the very least, avoid oxybenzone. Also beware of nanoparticles in sunscreen that cause DNA damage, inflammation, and mitochondrial dysfunction. The product may not say nanoparticles, so look for, and avoid, the following: fullerenes, micronized zinc oxide, nano zinc oxide, micronized titanium dioxide, and micronized quartz silica.

Tea

High levels of pesticides and plastics have been found in many popular brands of tea, including Lipton, Tetley, and Twinings. A 2018 CBC News investigation reported on the pesticide levels of

Lipton Pure Green Tea and Yellow Label Black Tea, Tetley, Twinings, No Name (an actual brand), Uncle Lee's Legends of China Green Tea, Red Rose, King Cole, and Signal teas. The investigation used an accredited lab and followed the same protocols as the Canadian Food Inspection Agency's for testing pesticide residue on the dry tea leaves.

The study found more than half the teas had residues above the legally acceptable limit. Eight of the ten also had other toxic chemicals. One brand contained over twenty-two different pesticides! The chemicals included endosulfan and monocrotophos, which are banned in many countries because of health risks and negative impacts on the environment. All toxins found were considered linked to everything from headaches and nausea to cancer, hormone disruption, and chronic illness.

The worst tea brand was Uncle Lee's Legends of China Green Tea, with over twenty different pesticides, including endosulfan. No Name teas contained over ten pesticides. King Cole had fewer pesticides but a wider variety of them. All other brands tested positive for pesticides. Only one brand tested negative for any toxins, and that was Red Rose Teas, which contained not one pesticide.

 Only one brand tested negative for any toxins, and that was Red Rose Teas, which contained not one pesticide.

According to the *Journal of Toxicology*, tea plants require a lot of rainfall and grow in acidic soil, which along with other contaminants includes aluminum and fluoride, which have been found in various teas due to the plants' absorption and deposition and concentration of the compounds in the leaves. Cadmium, lead, arsenic, and mercury are in the tea leaves before they ever reach the production stage and get put inside toxic tea bags. Levels of contaminants from teas made in China are worse if they are grown near coal-fired power plants. The use of coal in China has grown to 3.8 billion tons or about 47 percent of global consumption.

But pesticides and toxins aren't the only problem with tea. Plastics are present in most teas that are made with silken tea bags. A study from McGill University in Montreal, Quebec, Canada, found that silken tea bags release plastic particles when brewed, and most teas use these kinds of bags. Nathalie Tufenkji, a chemical engineering professor at McGill involved in the study, said, "We were expecting to see maybe a few particles of plastic … but we were really shocked when we saw the tea bags were releasing billions of plastic particles into a cup of tea." That equates to about $1/60^{th}$ of a gram of plastic in every cup of tea. You may want to consider learning how to brew your own with natural products and buy local, loose-leaf teas that do not use pesticides.

Toiletries

Two of the worst toxins are present in many toothpastes, mouthwashes, deodorants, antiperspirants, and other toiletry products: aluminum and triclosan. Aluminum in toothpastes and mouthwashes is ingested right into the body, and when coupled with fluoride found in these same products, it negatively affects the brain and has been linked to dementia, Alzheimer's, autism, and other brain disorders. Aluminum can also create "genomic instability" in breast cells and alter the environment of the breast enough to create oxidative damage, inflammatory responses, and cancer.

We know from a previous chapter of the dangers of fluoride, which is still found in most toothpaste and mouthwashes and can cause cancer, bone degradation, impaired IQ, kidney issues, receding gums, depressed cell growth, infertility, and heart disease.

Triclosan, which is also found in deodorants, antiperspirants, cleansers, hand sanitizers, facial tissues, laundry detergents, and antiseptics used on wounds, passes through the skin and interferes with hormone levels and functions. It is considered a known endocrine disrupter. It often appears on labels as TCS and is an antibacterial and antifungal agent, and it is even found in things like toys and various textiles. The FDA imposed a limited ban on this toxin from being used in many antibacterial soaps in 2016, but it is still found in many products, including shaving cream. Nontoiletry items that contain triclosan include credit cards, clothing, trash cans, and cutting boards. Scientific studies by Julie Gosse, Ph.D., a biochemist from the University of Maine, show that this chemical can cause harmful effects on the mitochondria and negatively affects reproduction.

Antibacterial soaps are very popular, especially during the COVID-19 pandemic, but did you know they contain triclosan, an additive that has been shown to affect hormones in animals and could harm the immune system.

Antibacterial soap contains triclosan, but its real negative effects come from its disruption of the body's delicately balanced microbial system. Regular soap and water work just as well to prevent infections, and they don't mess up our microbiome.

Shampoos and conditioners meant to clean and strengthen the hair include several "red list" toxins such as diethanolamine (DEA), an emulsifying agent that can increase the potential for cancer, liver and kidney damage, and brain development and memory. Many shampoos include formaldehyde; many consumers don't realize that the longer they keep the shampoo in the shower, the more cancer-causing formaldehyde is released.

Air fresheners kept in bathrooms or other parts of the home make the room smell great but release industrial chemicals that alter DNA. This includes gels, sprays, and plug-ins, which contain hazardous substances like formaldehyde, which cause eye and throat irritation, dizziness, wheezing, and coughing and contribute to lung damage, asthma, tumor growth, and hormone disruptions. If you think burning scented candles is a better idea, be aware that scented products can contain benzene and toluene, both of which are carcinogenic. Chemicals released from burning candles contribute to allergic reactions, asthma, and respiratory problems because the chemicals react to ozone in the air to create secondary pollutants. These secondary pollutants can wreak havoc on your endocrine system and your nervous system. Simmering essential oils in a diffuser is a better alternative.

The EWG's Dirty Dozen for People and Their Pets

The Environmental Working Group (EWG) published a list in 2013 of the dirty dozen endocrine disruptors. These toxins affect both humans and animals, so our pets are exposed to them from the same products we are:

1. Bisphenol A (BPA): Found in hard plastic water bottles, baby bottles, consumer electronics, sports equipment, CDs, DVDs, medical and dental devices, eyeglass lenses, and thermal paper, including store receipts.

2. Dioxins: Environmental pollutants created during industrial processes that disrupt male and female sex hormone signaling.

3. Atrazine: A herbicide used on corn that is also a xeno-estrogen, which can turn male frogs into females by af-

Mothballs

Many of us grew up with grandparents and parents who stored clothing in the closet or attic with mothballs to keep them clean and free of bugs. It was once common to open a closet and smell the distinct chemical smell. But these little balls of toxins contained almost 100 percent naphthalene or paradichlorobenzene, which are known to cause headaches, sore throats, coughing, nausea, eye irritation, even anemia, and with more long-term exposure, cancer, liver, and kidney damage.

If you have any mothballs hanging around the home, get rid of them. There are better ways to store clothing for the season that will repel bugs and not poison everyone in the house in the process.

fecting sex hormones. Also causes breast tumors and prostate problems in pets.

4. Phthalates: Found in PVC plastics and in air fresheners, candles, incense, children's and pet toys from China, nail polish, and perfumes. Inhaled into lungs or absorbed into skin and can end up in our bloodstreams. For pets, avoid plastic food containers.

5. Perchlorate: A chemical used in rocket fuel, fireworks, explosives, road flares, produce, and milk. It can disrupt the actions of iodine and negatively affect the thyroid hormones.

6. Fire retardants: The polybrominated diphenyl ethers (PBDEs) used are linked to a steep rise in hyperthyroidism in cats.

7. Lead: Damaging to human and pet organ systems in the body.

8. Arsenic: Same as lead.

9. Mercury: A naturally occurring yet toxic substance that is released through the burning of coal and finds its way into the seafood chain, including pet foods. Sar-

dines in a can are an exception, as they do not live long
enough to ingest a lot of toxins that then get consumed
by humans or pets.

10. Perfluorinated chemicals (PFCs): Used in nonstick
    cookware, textiles, rubber, plastic, and carpet treated
    with water-resistant coating.

11. Organophosphate pesticides: The dangers to humans
    have already been documented, but for pets, this
    danger can come from chemical pest-repellent collars
    and sprays in the house and yard. To offset this, use
    natural pest repellents on pets.

12. Glycol ethers: Chemicals found in solvents and used
    as stabilizers for personal and household products
    such as dyes, inks, paints, perfumes, and cosmetics, in-
    cluding aerosols that can also be harmful to pets in the
    home.

 Because pets are much smaller than people, these
chemicals get into their bodies in larger amounts.

These chemicals and the products that include them can increase or decrease specific hormones, disrupt their activities, bind
with other hormones, change one hormone into another, and even
act as if they are actual hormones in the body. All have the potential to do great harm to the body and every system that hormones
regulate.

Because pets are much smaller than people, these chemicals
get into their bodies in larger amounts. They live closer to the floor
or outside to the ground, where they can ingest household dust,
toxic cleaners, and pesticides and herbicides out on the lawn or garden. Think of the risk these toxins pose to children. Then think of
how small pets are compared to children, and you can see the
dangers of exposure to our four-legged friends.

Yoga Mats/Gym Mats

Mats used in yoga and gym classes can harbor tons of germs
from other users, but even the mats you buy for your own use come
with toxins such as polyvinyl chloride (PVC), which is a known carcinogen. Brand-new mats give off gas that is inhaled the first several

times they are used. When the mat is disposed of, the gas leaches into the ground and remains a hazard to the environment for decades. There are natural alternatives made of hemp and cloth, even recycled wet suit materials.

Plastic Food?

As if having bits of plastic in your food wasn't bad enough, now the Defense Advanced Research Projects Agency (DARPA) is working with Iowa State University and other entities, such as the American Institute of Chemical Engineers' RAPID Institute and Sandia National Laboratories, to convert paper waste into sugars and plastic into fatty acids and fatty alcohols. This would be made into actual food to provide short-term nourishment for troops on extended missions.

Should plastic food be a hit, it will no doubt appear on store shelves soon for a cheap food choice. These researchers see it as a way to help reduce plastic pollution in the environment by repurposing it as consumables, but one has to wonder who would want to eat something like this? Plastics are biodegradable, but it is a very slow process, so the idea is to increase the "oxo-degradation" of plastics into fatty compounds by exposing them to ultrahigh temperatures. Then the cooled result is used to grow yeast and bacteria into single-cell proteins that can become foodstuffs.

Maybe this could be a win for the environment, and perhaps in this form, plastics do not pose as much of a danger to the human body as when they are involuntarily consumed as microplastics in our food or in the wrapping and containers we buy and save our food in, but selling the public on it might be a bit of a stretch.

BIG PHARMA: POWER AND THE POISON PILL

In our Big Pharma world, those who seek natural alternatives, or naturopathic medicine, instead of relying on allopathic medicine are often censored, banned, shamed, and ridiculed, yet are we, the pill-popping generations, healthier than past generations? No. A resounding no. In *Population Control: How Corporate Owners Are Killing Us*, author and investigative journalist Jim Marrs states, "There is no logical reason—apart from the profits enjoyed by the elites at the head of pharmaceutical companies—why our health care system should focus exclusively on these allopathic 'cures.'" He later says that the cozy relationship between the Food and Drug Administration (FDA), Big Pharma, and Congress allows deadly medicines to make their way onto the market and to keep over-the-counter drugs that themselves can be dangerous on store shelves. It also keeps many drug prices beyond the reach of the average working American citizen.

 We are being poisoned and robbed by Big Pharma, all because we have been led to believe it can cure what ails us.

We are being poisoned and robbed by Big Pharma, all because we have been led to believe it can cure what ails us. But we, the consumers, are equally guilty for wanting the quick and easy

More and more, people are taking prescription medications for just about everything, and there's a good chance they do not need all the pills and liquids they are swallowing these days.

fix. Escaping the grip of Big Pharma is becoming more challenging as it lobbies our politicians, creates hundreds of new medications, buys tons of advertising to push them (many with pretty pictures and happy people dancing about to joyful music!), and gets in bed with social media companies and heavy hitters such as Bill Gates to blanket the population with their singular message: "Only we can save you. Take more pills/meds/vaccines."

Lobby Power

Our political system includes corporations and entities that put pressure on our elected officials to vote for or create policies that benefit their bottom line. This often comes in the form of monetary donations to said elected officials. In the past, the oil and gas, tobacco, and auto industries led the lobbying charge. Today, it's the pharmaceutical lobby to the tune of $4.45 billion over the last twenty-two years. Big Pharma has, according to the Center for Responsive Politics' political contribution and lobbying spending list from January 1998 to March 2020, "far outpaced all other industries in lobbying spending." The focus of all this lobbying has been "resisting government-run health care, ensuring a quicker approval process for drugs and products entering the market, and strengthening intellectual property protections."

As of March 2020, there were 1,227 pharma/health product lobbyists in the United States, and—get this—62 percent were former government employees. That's right. Former congressional representatives often become the next drug lobbyists. The pay is much better. If you're wondering which industries are second and third behind Big Pharma, it's the insurance industry and electric utilities. The insurance industry is often tied directly to the lobbying of pharmaceutical companies, and it has become involved in the legislative process to influence new regulations and curb any enforcement of laws that hurt its bottom line.

A January 2019 article by Susan Scutti for CNN Health, "Big Pharma Spends Record Millions on Lobbying Amid Pressure to Lower Drug Prices," reports that a record amount was spent on behalf of the Pharmaceutical Research and Manufacturers of America (PhRMA), which represents the biggies in the industry, including Pfizer, Merck, Johnson & Johnson, and Gilead Sciences. OpenSecrets, a nonpartisan, independent research group that tracks money in U.S. politics, stated there was over $194,000,000 extra used for lobbying above and beyond what PhRMA officially disclosed for 2018.

 The goal isn't just to prop up a politician one time but rather to build loyalty and create an obligation by that politician to scratch the backs of those who scratched them first.

Often, these cash donations target lawmakers who are more vulnerable or in need of an extra boost around election time. Political action committees (PACs) made up of pharma employees and associated trade groups will focus on politicians who are up for reelection in the current or next election cycle and infuse their coffers with extra cash. In 2019, the Kaiser Health News's "Pharma Cash to Congress" database showed a hefty $845,000 donated to thirty senators up for reelection in the following year. The goal isn't just to prop up a politician one time but rather to build loyalty and create an obligation by that politician to scratch the backs of those who scratched them first. PhRMA spokesperson Holly Campbell is quoted as saying PhRMA supports candidates of both political parties who "support innovation and patient access to medicines." Wondering who the biggest beneficiaries of the year 2019 were? Senator Chris Coons, a Democrat from Delaware, received $103,000 while Senator Thom Tillis, a Republican from North Carolina, was a close second at $102,000. And that was just for the first six months of that year. Both parties are equally guilty of taking in Big Pharma dollars in return for legislating for the industry's benefit. In the past decade, Congress members from both sides of the political aisle have received about $81,000,000 from sixty-eight different Big Pharma PACs. Who wins? They do. Who loses? The consumer, no doubt.

However, it isn't always votes these Big Pharma companies want with their lobbying dollars. Often, it's influence in creating upcoming legislation and controlling aspects of industry regulation, so they are, in a sense, also buying access to the men and women who can deliver the goods.

It isn't just politicians receiving big money donations and gifts from Big Pharma. Doctors receive payments, bonuses, and gifts from pharmaceutical companies for pushing certain drugs or vaccines (more on those later). Ever see a drug rep come into your doctor's waiting room? They are usually dressed to the nines and have a huge briefcase or bag filled with samples and goodies. Notice how you wait an hour to see your doctor, but the drug rep is welcomed back immediately? Money buys not just influence but access, and no doubt, the drugs they are pushing will end up being recommended to you at some point.

ProPublica, a nonprofit newsroom that is devoted to investigating abuses of power, publishes *Dollars for Docs*, which tracks pharma company payments to doctors. Federal law requires that drug and medical device companies make their payments public, and *Dollars for Docs* has become the essential tool for patients who want to know who is influencing their doctors and with how much money. Their analysis of these payments concluded that in many cases, doctors were prescribing the drugs from companies they received payments from, and many top prescribers of brand-name, expensive new drugs did have relationships with the pharma companies. ProPublica also found that the more money the doctors received, the more brand-name medications they prescribed. Consumers can go to ProPublica's website, propublica.org, to learn more by using its "Prescriber Checkup" tool that shows which doctors received a payment related to the top fifty brand-name drugs in Medicare and other pertinent data.

Industry Corruption

Rather than direct lobbying, Big Pharma also influences nonprofit patient groups and organizations by providing them with funding that is not disclosed or is underreported. Patient-oriented and patient-advocacy groups alike are supposed to be there to help consumers, but some take money from the "enemy" and allow Big

Pharma to influence their operations and research decisions. In "The Pharma Façade: More Hidden Money Going to Nonprofits" from Mercola.com, we learn that drug companies often finance advocacy groups and disease-awareness campaigns, while support for patients only gets about 5.9 percent of their payments. The priority of the drug industry is not on patients but on commercially driven PR and profits, with cancer and diabetes getting the most funding since they are considered to have the most "commercial potential." The priorities of the pharma industry are based not on helping people to live but on which conditions have the most commercial potential to be exploited.

The article goes on to state, "What's more, the financial relationships between patient organizations and drug companies are often not transparent, with the pharmaceutical industry under-reporting payments." Both the donors and the recipients were guilty of underreporting payments. "Existing donor and recipient disclosure systems cannot manage potential conflicts of interest associated with industry payments," and the extent of this underreporting is unknown without any public list.

In "Corruption in Global Health Care May Exceed $1 Trillion," Dr. Joseph Mercola looks at the dishonesty, fraud, and corruption in the health care system's academic and research communities, which are rarely addressed or condemned since Big Pharma has such a stranglehold on our political system. The less transparent health care systems are, the more they are open to corruption, and "drug research is beset by an astoundingly high incidence rate of scientific misconduct, with 72% of retracted drug studies due to things like data falsification, fabrication, unethical conduct, and plagiarism." Scientific fraud is a big deal but one the consumer rarely hears about, and the result is a plethora of drugs and medications pushed to consumers that may be unsafe or, worse, deadly.

One such example given in the above article was noted in the April 2013 *American Journal of Medicine*, referring to recent allegations of fraud committed by a researcher named Dr. Don Poldermans, who wrote a fraudulent beta blocker study that was linked to the deaths of 800,000 Europeans. Poldermans's paper was highly influential to many researchers who came after him and was instrumental in establishing the use of beta blockers in noncardiac surgery patients as standard care. Poldermans was also the chairman of the committee that drafted this guideline!

Beta blockers are a mainstay of treating high blood pressure, with millions of people taking them thinking they are improving

The Centers for Disease Control and Prevention (CDC) is a government agency charged with the protection of public health in America. There is concern that accepting funding from private interest groups could compromise that mission.

their health. Yet the integrity of the main study that established this drug for such use was fraudulent.

There are suspicions that the Centers for Disease Control and Prevention (CDC) is guilty of accepting funding from special interests. So, the main organization consumers look to for advice on everything from sunscreens to vaccines takes money from the companies that make those products. Between 2014 and 2018 alone, watchdog groups such as the U.S. Right to Know, Public Citizen, and Project on Government Oversight report that the CDC took over $79,000,000 from drug companies and commercial manufacturers. Since its inception in 1995, it has accepted over $161.000,000 from private corporations. If this isn't a massive conflict of interest, it's hard to find what is. No doubt, these financial donations impacted the CDC's choices on what it tells the public in terms of health recommendations. More on this in the vaccine chapter.

In a 2019 paper published in *The Lancet* by Dr. Patricia Garcia, affiliate professor of global health at Cayetano Heredia University, she states that corruption in the form of dishonesty and fraud in the health care system is one of the most important barriers to implementing universal health coverage. "It is estimated that, each year, corruption takes the lives of at least 140,000 children, worsens antimicrobial resistance, and undermines all of our efforts to control communicable and non-communicable diseases. Corruption is an ignored pandemic."

It is a pandemic to the tune of hundreds of billions of dollars each year of the $7 trillion the United States spends on health services. Of that, it is estimated 10 to 25 percent is lost directly through corruption.

This fraud extends into actual clinical trials of drugs, with pharma companies funding many of the studies that get published in medical journals. Pharmaceutical companies may also design an entire study and handpick the results. According to "How Big Pharma Influences Trial Results" for Drugwatch.com, *"The New England Journal of Medicine*—one of the most prestigious medical journals in the world—published 73 studies of new drugs. Of those studies, a pharmaceutical company funded 60; 50 had drug company employees among the authors and 37 lead researchers had accepted money from a drug company, according to a review conducted by the *Washington Post*." These medical journals are often the first to announce a new drug to market and are trusted by consumers because who on Earth would think a medical journal would lie or distort the truth? But they do, not only by publishing clearly tainted and sponsored studies but also by not being transparent about it until a watchdog group takes them on.

A report from Johns Hopkins University found that the number of clinical trials funded by Big Pharma increased every year since 2006, while funding by the National Institutes of Health (NIH), a more independent agency that took far less pharma money, fell. In 2014 alone, Big Pharma financed 6,550 trials, while the NIH funded 1,048, according to a study by Stephen Ehrhardt published in *JAMA: The Journal of the American Medical Association*. This proves that Big Pharma has control over so much more than just our politicians and doctors. They hold sway over the very medical studies that determine whether a drug should get a passing grade and be taken to market. Those drugs, as we will see soon, have often resulted in deaths and major lawsuits.

Who can you trust when it comes to your health? Clearly not who you thought.

Toxic Pills

While the War on Drugs focused on street and illegal drugs, Americans were doping up far worse on pills and treatments that were proving to be toxic to their health. Ironic, as they were taking these medications to save their lives or improve their health. But no pill comes without a long list of side effects and chemicals that are toxic to the delicate systems of the human body. This book would have to be 50,000 pages to talk about all the toxic drugs out there, so we will focus on some of them. By the end of this chapter, it will be clear that America is overdrugged, and it has only served to make us all sicker, more tired, and full of chronic illnesses. It has created a health crisis by taking the focus off of lifestyle changes and natural methods of increasing health and well-being.

 By the end of this chapter, it will be clear that America is overdrugged, and it has only served to make us all sicker, more tired, and full of chronic illnesses.

Side Effects

The most toxic element of taking pharmaceuticals, even over-the-counter (OTC) drugs, is the long list of side effects that can cause as much grief, if not more, than the original reason for taking the meds. Sadly, few doctors take the time to go over side effects when prescribing drugs, leaving that job to pharmacists, but usually only if the consumer asks. Yes, each medication lists some of the side effects on the box or insert, but many consumers have found that these lists are not complete or always honest.

Many consumers in group and support forums complain of side effects to drugs that their doctors insist don't exist. Is this arrogance on the part of the medical profession? Further arrogance is to insist that someone is not experiencing something that they did not experience before taking a particular medication. In a 2017 *Washington Post* article by Janice Neumann, "Doctors Often Don't Tell You about Drug Side Effects and That's a Problem," the reporter states that it may be that drug companies don't inform or warn the doctors sufficiently about side effects. Adriane Fugh-Berman, in the Department of Pharmacology and Physiology at Georgetown University Medical Center, adds, "Doctors often prescribe irrationally, pushing branded over generic drugs, new over old, and suggesting medications for diseases invented or exaggerated by drug companies." She views this as a bad situation, where people in control of prescribing the drugs often know the least about the drugs.

Have you seen the commercials that tell you to ask your doctor about whether a certain drug can help you? This campaign banks on your doing just that because the pharma companies make a bundle for selling these products, whether you need them or not.

Lawsuits

Taking drugs with side effects often results in damages and injuries that lead to major lawsuits and recalls. In April 2020, the FDA pulled the popular heartburn drug Zantac from the market because of links to liver cancer, kidney cancer, stomach cancer, colon cancer, prostate cancer, pancreatic cancer, digestive tract cancers, and other cancers. Several lawsuits are ongoing against the makers of Zantac, Sanofi and Boehringer Ingelheim, to hold them accountable for reckless disregard for human health and putting profits over the health of people.

In the past few years, the makers of the antidepressant Paxil paid out almost $3 billion in fines for illegally persuading doctors to give the drugs to children and teens despite warnings that it could trigger suicidal behaviors. The diabetes drug Actos's label didn't include a warning of the risk of bladder cancer on the package, resulting in a payout in settlements to the tune of $2.34 billion. In 2012, Bayer paid out $110,000,000 to settle allegations that consumers experienced fatal blood clots when using Yasmin, a popular birth control pill. Earlier, the makers of the painkiller Vioxx paid out $4.85 billion to settle 50,000 claims of related heart attacks and strokes. This is just a small sampling of the lawsuits that have been settled against Big Pharma.

In many cases, because the Supreme Court has never held Big Pharma liable for injuries and deaths, people might not have been able to turn to a lawsuit for some recompense from damage done by drugs. But recently, criminal suits and major class-action lawsuits have succeeded in calling the drugmakers to task. The power does lie with the consumers, though, to be self-advocates.

Researchers from Brigham and Women's Hospital and Harvard Medical School found that even as more drugs get to market, the percentage of necessary drugs decreases. Since the mid-1990s, almost 85 percent of new drugs do not offer any clinical advantages to the consumers.

A massive number of lawsuits have been leveled at the makers of opioids, deadly painkillers that have taken the lives of over a quarter-million Americans and made many more addicts. There are even federal trials against pharmacy chain stores that sold the drugs in the works as well as criminal and civil cases that could cost the makers of opioids billions of dollars when the dust is settled. Johnson & Johnson has already settled big lawsuits, including a half-billion-dollar settlement in an Oklahoma lawsuit where the state's attorney general called the drugmaker a "drug kingpin organization."

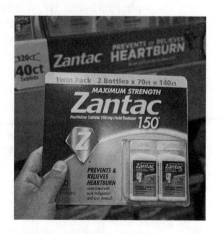

A whistleblower lawsuit was filed against Roche, the maker of the flu medication Tamiflu, alleging that Roche made false claims and put forth fraudulent studies that resulted in the U.S. government buying over $1.5 billion of the drug for future outbreaks. The lawsuit claims Roche knew the

One of many drugs to be the center of a lawsuit, the heartburn drug Zantac was linked to liver, kidney, stomach, colon, and other cancers in consumers.

drug was ineffective and that the company marketed it anyway to make profits at the expense of consumers and users. This lawsuit could cost the company over $4 billion thanks to the False Claims Act, which triples the damages made as well as offer civil penalties.

GlaxoSmithKline holds the distinction of paying out one of the largest health care fraud settlements in history. It paid out a criminal fine of $956,814,400 with a forfeiture of $43,185,600 and $2 billion to resolve civil liabilities under the False Claims Act for the drugs Paxil, Wellbutrin, and Avandia. Just about every Big Pharma company has been hit with lawsuits over not reporting side effects or committing fraudulent claims, including Pfizer for Lyrica and Zyvox, Eli Lilly for the antipsychotic Zyprexa, and AstraZeneca for Seroquel. Sometimes the violations are off-label promotions, Medicare/fraud kickbacks, monopoly practices, and poor manufacturing practices. Sadly, these lawsuits have not stopped pharmaceutical companies from continuing these practices. The settlements are often just a drop in the bucket compared to the profits they make.

 Every now and then, a pharma executive goes to jail.

Every now and then, a pharma executive goes to jail. This happened in early 2020 when John Kapoor, a former billionaire and pharma executive, was sentenced to over five years in prison as the result of a U.S. District Court criminal trial, the first successful prosecution of a pharma executive tied to the opioid epidemic. Kapoor, founder of Insys Therapeutics, was sentenced to sixty-six months in prison, less than the fifteen years prosecutors wanted him to serve for his role in aggressively marketing a potent opioid, Subsys. Kapoor and four other executives were found guilty of criminal conspiracy to bribe doctors to prescribe Subsys, an oral fentanyl spray, even to patients who did not request or need it, then lied to insurance companies to get them to cover the cost.

Prescription and OTC drugs have their place and can be lifesavers in many situations. But they must be researched for side effects, and one of the best places to find out about side effects not normally listed on the insert or box is consumer forums. People who have used the drugs often report things of which doctors are not even aware. Consumers can and should press both doctors and pharmacists for a list of side effects and ask about pending lawsuits, too. It always comes down to buyer, or user, beware.

OTC Painkillers

The following is only a small sampling of the toxicity of specific drugs and OTC medications, but it gives a chilling look at how common dangerous side effects and toxic ingredients are in the pills and treatments we take to ease our pain and maybe even save our lives.

Most people take OTC painkillers such as aspirin, acetaminophen, and ibuprofen, unaware of the dangers each carries. Aspirin has been a mainstay for decades for headaches, sore throats, and any kind of physical pain and has been a part of a protocol for those who want to prevent heart attacks and strokes. Doctors have prescribed millions of people baby aspirin once a day for preventing heart attacks and strokes, but taking an aspirin daily has serious consequences, including a higher risk of internal bleeding and stomach ulcer formation.

Those side effects include the spread and progression of cancer cell growth in adults over the age of 65, according to a new study in the *Journal of the National Cancer Institute*. In the study, researchers looked at daily use of 100 mg of aspirin in 2,411 people over 65 years of age in the United States and 16,703 in Australia 70 years and older. No participants had any signs or history of heart disease, dementia, or other physical illnesses. Among the group taking the daily aspirin, there were 981 cancer events. Only 952 events occurred in the group taking a placebo. Deaths were higher among those with advanced cases of cancer who took the aspirin. The study concluded a link between taking daily aspirin and the occurrence of metastasis or the progression of cancer to a stage 4 diagnosis and found that patients with stage 3 cancer even had a higher risk of dying.

Another large study, the Aspirin in Reducing Events in the Elderly (AS-PREE), had previously found that the rates of stroke, heart disease, and dementia were the same in people taking daily aspirin as in those who did not take it. However, increased risk of bleeding happened more frequently

People have been led to believe that taking a low dosage of aspirin once a night can stave off a heart attack. The truth is, there is no solid evidence for this, but it does increase the risk of internal bleeding.

even with a small daily amount of aspirin, and that rate increased with higher amounts, leading to brain, stomach, and intestinal bleeding risks. Another study presented at the Annual Congress of the European League against Rheumatism found that nonsteroidal anti-inflammatory drugs (NSAIDs) such as aspirin and ibuprofen could cause heart attacks.

The data in the study demonstrated an increase in the risk of death and heart attacks enough to challenge the safety of using them even in the short term. Ten thousand patients were analyzed in the study, all over fifty years of age, who had taken NSAIDs in the past and had been previously diagnosed with ischemic heart disease, diabetes mellitus, and/or hypertension. The study concluded that heart failure risk nearly doubled with NSAIDs. There was also a marked increase in upper gastrointestinal issues and a one-third increase in major vascular events by a COX-2 inhibitor. Ibuprofen significantly increased major coronary events but did not increase major vascular events.

Another study, published in the *European Heart Journal— Cardiovascular Pharmacotherapy*, showed a significant 31 percent increased risk of heart attacks with NSAID use. NSAIDs work by blocking proinflammatory enzymes called COX-1 and COX-2, which suppress the body's own natural inflammatory response. This rids the pain symptoms, but it doesn't treat the cause, and in the end, it can be more dangerous than suffering through a headache or sore throat. NSAIDs not only raise the risk of blood clots by affecting platelet aggregation, but they also constrict the arteries, increase fluid retention, and raise blood pressure, all of which lead to higher risks for the very diseases that they are prescribed to help decrease.

When someone has a heart attack, the first action taken is to administer aspirin or an NSAID such as Bufferin, Excedrin, or baby aspirin. But in the longer term, finding alternatives to these and other NSAIDs such as Advil, Motrin, Aleve, Nuprin, Actron, and Orudis might be better for your heart. With these products sold everywhere and convenient and cheap, it is a hard addiction to break but one that might be worth it.

All analgesics, or pain medications, seem to have some negative impact on the kidneys. According to the National Kidney Foundation, as many as 3 to 5 percent of new chronic kidney failure cases every year are related to the overuse of pain meds, and once the disease occurs, continued use worsens the condition. Long-term use of these meds causes harmful effects to the kidney tissue and structures. It is highly recommended to use the lowest dosage possible when treating pain. If you have existing kidney problems, avoid aspirin, ibuprofen, and acetaminophen altogether.

General side effects of taking ibuprofen specifically include:

▲ Stomach pain

▲ Heartburn

▲ Gas

▲ Vomiting

▲ Nausea

▲ Constipation

▲ Diarrhea

▲ Increased blood pressure

▲ Dehydration

▲ Fluid retention

▲ Dizziness

▲ Less frequent urination

▲ Ulcers and stomach and intestinal bleeding

▲ Kidney damage

▲ Liver failure (very rare but possible)

Tylenol™

If aspirin and NSAIDs are toxic to the body, one might choose acetaminophen instead, but that might be an even worse choice. The most popular, Tylenol™, has many dangerous side effects and a high risk of overdose rates because people pop these pain meds like candy, thinking they are totally safe. Taking this drug with alcohol can damage the liver, as can taking it on an empty stomach. Because of its effects on the liver, it results in approximately 56,000 emergency room visits annually in the United States, as well as 100,000 calls to poison control, 26,000 hospitalizations, and almost 500 deaths.

Your Brain on Drugs

There are several prescription and OTC drugs that may have adverse effects on your brain. These run the gamut from antihistamines to painkillers, which can sedate the brain, lower brain function, and in older people be a real risk for falling or causing additional problems to an existing illness. Use with caution, especially if you are older or are administering to children.

Medications that cause sedation such as nighttime pain relievers and antihistamines such as Benadryl (overprescribed to young children and should not be!); vertigo medications such as meclizine, which is sold as Dramamine nondrowsy and Antivert; medications for an overactive bladder; nerve pain medications, especially tricyclics, which are an older class of antidepressants; and muscle relaxers.

Overuse of the above can cause problems with memory retention and normal focus and functioning by making you feel dizzy, tired, or unsteady. Those who may be dealing with dementia and Alzheimer's may want to avoid them all and find safer alternatives, and younger people might avoid these brain toxins and protect their brain cells longer, pushing off dementia and Alzheimer's by years.

The makers of Tylenol™, Johnson & Johnson, are allowed by the FDA to sell this dangerous drug over the counter without any black-box warning about how it can even damage the liver at standard doses. This is because when the pills are metabolized, they release a toxin called N-acetyl-p-benzoquinone imine. The risk of kidney and liver damage increases for children and infants. It is disturbing that such a drug can be purchased by anyone without a prescription or a warning.

But it gets worse. Medications that contain acetaminophen have been found to trigger negative emotions and negatively affect mental and emotional wellness. A 2015 study published in the journal *Psychological Science* gave half the participants a placebo and the other half a dose of 1,000 mg of acetaminophen. After an hour passed, all participants took the same test, viewing photos designed to cause either a positive or negative response. Researchers found the acetaminophen group had a dampened or dulled reaction with both highs and lows compared to the placebo group. A 2016 study published in *Social Cognitive and Affective Neuroscience* involved eighty college students, half given 1,000 mg of acetaminophen and half given a placebo. The students were asked to read stories of people going through pain and to rate the pain level of each story. The result showed much lower pain ratings given for people in the stories compared with the placebo group, indicating that the drug could be dulling the ability to feel and experience empathy and compassion.

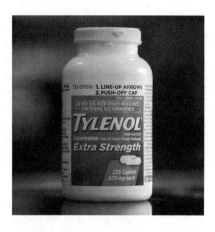

Tylenol™ might not be as safe a pain reliever as you think. Every year, hundreds of Americans die from overdosing on this common, over-the-counter drug.

Another study involved 189 college students, half given 1,000 mg and half a placebo, and this time, they were asked to rate the risk level of activities like bungee jumping, skydiving, and walking alone in an unsafe area. The result showed that the students who took the acetaminophen rated dangerous activities as less risky than the placebo group. In a 2020 article titled "Study Links Tylenol™ Consumption with Risk Taking" for GreenMedInfo, researchers stated that so many people taking Tylenol™ was cause for alarm. "The psychological side effects of acetaminophen, aka Tylenol™, continue to mount, with research showing users are more likely to make risky decisions. When coupled with past research linking this supposedly safe pain reliever to blunted empathy and emotions, the widespread social effects on society could be immense."

Other studies in the last few years have shown that taking Tylenol™ or any other acetaminophen clouds judgment; increases the rate of making errors; reduces the distress associated with making errors; affects feelings of social rejection, uncertainty, and the evaluative process; and blocks neurological responses associated with both physical and emotional distress. They also water down and blunt feelings of positivity and negativity. Basically, this class of drugs, which is sold everywhere and taken every day, can be making us become zombies, dulling our feelings, and lessening our compassion toward others.

Dominik Mischkowski, a pain researcher from Ohio University, uncovered some very disturbing data in his own study with paracetamol, a generic acetaminophen. In "The Medications That Change Who We Are" by Zaria Gorvett in the January 8, 2020, issue of BBC Online, Mischkowski wondered if painkillers might be making it harder for people to experience empathy, especially since the same areas of the brain that become active when we experience positive empathy are active when we experience pain. He recruited college students and gave half the group a placebo and half 1,000 milligrams of paracetamol. They were then exposed to stories about other people having uplifting experiences and gauged their responses. The study showed, as others did, that the drug group had a significant reduction in the ability to feel positive empathy, so much so that he stated, "To be honest, this line of research is really the most worrisome that I've ever conducted. Especially because I'm well aware of the numbers involved. When you give somebody a drug, you don't just give it to a person—you give it to a social system. And we really don't understand the effects of these medications in the broader context."

 The study showed ... that the drug group had a significant reduction in the ability to feel positive empathy....

That same article documented the negative effects of other drugs on human behavior, such as statins, antidepressants, and asthma medications, all of which have been shown in studies to change our behaviors, shape our social relationships for better or worse, and affect our brains in negative ways, sometimes fearful. "They've been linked to road rage, pathological gambling, and complicated acts of fraud," reporter Gorvett stated. Chilling.

A 2019 study by researchers at Johns Hopkins University analyzed umbilical cord samples in pregnant women to measure levels of acetaminophen (Tylenol). They then compared it to the number of reported cases of ADHD, autism, or developmental disabilities and found that there were twice as many of these abnormalities in children whose mothers had taken Tylenol. However, it should be noted that other scientists who read the study found that the Johns Hopkins researchers did not consider other factors, including family history, environment, and genetics.

There are other risks for pregnant women using acetaminophen, such as a higher rate of pre-eclampsia and thromboembolic diseases and a higher risk of preterm birth if used in the third trimester. It has also been reported that taking the drug longer than four weeks during the first and second trimesters moderately increases the risk of undescended testicles in boys.

In terms of autism, the increased use of this drug in children after they are born is also a factor, and many argue that the increased risk of autism seen following childhood vaccines may be due to the inappropriate use of acetaminophen administered immediately after the shots and not just the vaccines themselves (which also contain toxins that are linked to autism).

The Opioid Epidemic

One of the most widely abused prescription drugs is opioids, which are incredibly addictive and responsible for rising death rates. According to CNN, the United States is in the throes of an opioid epidemic with upward of 10,300,000 people aged twelve and older misusing the potent painkiller. This includes 9,900,000 prescription users and another 808,000 who use heroin. In 2018 alone, there were 46,802 overdose deaths due to opioids. The only bright side to this deadly epidemic is that sales of prescriptions peaked in 2011, with about 240 billion milligrams of morphine prescribed, but fell dramatically in 2018 by 29.2 billion morphine milligram equivalents.

Yet these drugs remain a clear and present danger to those who get hooked on them. What makes them so powerful is the way the brain becomes used to the feelings these drugs induce so that more is needed to achieve the same high. Opioids like morphine and codeine come from the poppy plants grown in Asia, South America, and Central America, and heroin is simply an illegal, synthesized version of morphine.

Opioids like fentanyl, which surpassed heroin and oxycodone as the leading cause of overdose deaths in 2016, were originally developed to use in surgery and to alleviate the severe pain associated with terminal cancer patients. Fentanyl is 100 times more powerful than morphine, and a tiny amount can be deadly. Fentanyl is often added to heroin to make it even more potent, and people who use the drug don't know they are taking the deadlier fentanyl-laced product. One of the biggest manufacturers of heroin laced with fentanyl is Mexico, and it goes by several street names such as China White, Great Bear, Poison, and Murder 8.

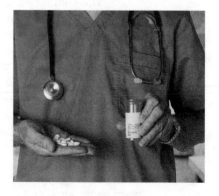

Doctors in the United States have been overprescribing opioids for pain, which has contributed to a crisis in addictions and overdose deaths.

The high from fentanyl doesn't last long, but the toxic after-effects can include reduced blood pressure, fainting, slowed respiration, coma, and even death. A high price to pay for a short high.

Methadone is another opioid interestingly used to help heroin addicts kick the drug. Percocet and Percodan were popular drugs given to people getting their wisdom teeth pulled decades ago. These drugs all work the same way: they bind to receptors in the brain and spinal cord that disrupt pain signals, activating reward areas of the brain to release dopamine and create the euphoric high they are known and hunted down for.

In years past, OxyContin has become a Big Pharma profit dream as prescriptions rose with more people requesting the opioid and more doctors freely giving it out. Because of its highly addictive properties, it skyrocketed in use to the point where people were buying it off the dark web or black market. Purdue Pharma LP, the manufacturer of the powerful and potent opioid, agreed to plead guilty in October 2020 to criminal charges over its handling of OxyContin due to its proclivity for pushing the flow of medically questionable prescriptions and downplaying the addictive nature of the drug.

 With a ton of media attention and public outcry, more has been done to curb opioids than just about any other drug....

The company pleaded guilty to felony counts that violated federal antikickback laws and one charge of defrauding the United States and violating the Federal Food, Drug, and Cosmetic Act, but it was able to avoid paying penalties of over $8 billion, instead agreeing in the deal to pay $225 million toward a $2 billion criminal forfeiture, with the Justice Department foregoing the rest of the amount under certain restructuring circumstances. Suffice it to say that this is just a drop in the bucket of the profits Purdue made by creating millions of prescription drug addicts to its golden child, OxyContin.

These drugs are so easy to get addicted to because of the reward sensations, and over the last few years, major legislation has been introduced to increase research and treatment for opioid addicts and to regulate pain clinics by limiting the number of opioids doctors can dispense. With a ton of media attention and public outcry, more has been done to curb opioids than just about any other drug, legal or illegal, but it still manages to find its way to our borders and into the United States to create new addicts and new deaths.

Drugs and Children

If drugs are toxic to adults, imagine the damage they do to smaller bodies. Children are given everything from Tylenol™ to Benedryl to cough syrups without concern for what toxic buildup and side effects they might cause. Not only are OTC drugs being doled out on children, many of whom are completely without need of them and overmedicated by them, but we now have our kids taking prescription drugs for ADHD, allergies, chronic illnesses, stomach pain, gastroesophageal reflux disease (GERD), depression, and anxiety.

When it comes to OTC drugs, the focus is on pain meds and antihistamines, neither of which should ever be given to any child under the age of two and are not really recommended for those under the age of six. According to the American Association of Poison Control Centers, the top three products associated with deaths in children under the age of five are OTC cough and cold medications!

Some of the toxic ingredients in these medications include guaifenesin, dextromethorphan, acetaminophen, and food colorings added to make syrups palatable, not to mention sugars. Dosages on the product are often general and don't consider any existing conditions the child may have. Worse still, most of these medications only cover up the symptoms for a short while and then need to be administered again.

Another new concern is diet medications being given to children to help them reduce their weight. Many parents don't even tell their child's doctor out of shame, and then, they run the risk of the child being given a drug that counteracts with the diet drugs or supplements.

Exercise extreme caution and restraint giving children under the age of six medications. Cough and cold medication and antihistamine overdoses can cause everything from heart problems and seizures to death in the very young.

Two of the biggest issues with any drug for children is accidental overdose and overstimulation or oversedation. But as more children of all ages fall to chronic illnesses such as diabetes, depression, anxiety, GERD, and obesity, the number of prescriptions written for them will increase in return. Our adult population is drugged up and overmedicated enough. We cannot allow the same to happen to our children.

We also overmedicate our elderly population, many of whom are taking more than five pills a day. Because many are getting less physical activity and eating less or maybe not as healthy, there is a drive to turn to the power of the pill. With rising rates of dementia, Alzheimer's, cancers, diabetes, and heart disease as people age, the number of pills they take rises as well, increasing risk of an earlier death, horrific side effects, and bigger profits for the pharmaceutical companies. With the elderly, it is often a dangerous loop of getting sick or injured, being medicated for it, then having side effects, and being medicated for those, all in a quest to cover symptoms while doing nothing to improve the quality of life.

Statins

Over 25 percent of Americans over the age of forty take a statin drug to lower their cholesterol. In 2019, the sales of one of the most widely prescribed statins, Lipitor, made Pfizer Pharmaceuticals over $2 billion. Yet statins have a laundry list of side effects and have failed to stop the oncoming train of heart disease. The debate over cholesterol is a complicated one: the result is the overprescribing of a drug that can indeed lower cholesterol levels but at a high price despite plenty of scientific research showing that heart disease is caused by a host of issues, including high blood pressure, lack of exercise, poor diet, poor sleep, inflammation, and stress.

One 2005 Columbia University study even found that people with the lowest total cholesterol and LDL cholesterol (the "bad" as opposed to HDL, the "good") levels were twice as likely to die within three years compared to people with the highest levels.

Statins are notorious for causing muscle stress, pain, fatigue, and weakness, including sharp cramping and tendonitis, which can lead to myositis (inflammation of the muscle) and rhabdomyolysis (severe muscle damage, cell death, kidney failure, possible death). Statins also inhibit the birth of new mitochondria, which are the energy "batteries" of cells and can even worsen symptoms in those who suffer from muscular dystrophy.

If you like to exercise, statins increase the possibility of muscle injury.

They also raise insulin levels in the blood. In a growing population suffering from insulin resistance and Type 2 diabetes, statins can trigger these diseases and increase the risk of new-onset diabetes to boot, especially for those people taking higher dosages of statins after a heart attack or surgery. They also cause brain fog, impaired concentration, memory loss, and mood changes and disturbances and increase the risk of obesity. Ironic, considering obesity contributes to heart disease, the very thing most people are prescribed statins to reduce the risk of.

So many seniors are put on statins, unaware of the cognitive impairment the drug can cause. This includes not just memory issues and lack of focus and concentration but also increased rates of depression. These symptoms are often swept aside by doctors as being the cause of natural age, but those who have stopped statins can attest to how well they feel once their brains are free of the drug.

 Statins contribute to liver damage from the rise in the liver's production of enzymes that cause inflammation, and they increase the levels of dangerous plaque in coronary arteries....

According to the *BMC Cancer* journal, prolonged use contributes to colorectal, bladder, and lung cancers. Statins contribute to liver damage from the rise in the liver's production of enzymes that cause inflammation, and they increase the levels of dangerous plaque in coronary arteries, thus raising the risk of a heart attack.

So much for taking a pill to help your heart. There are many natural ways to lower high cholesterol and protect the heart from disease. Specifically, many people have turned to red yeast rice, which has the same effects as lowering high cholesterol but can also come with milder side effects unless monitored and taken with CoQ10, which is depleted from the use of statins. Fiber, especially psyllium husk, can do the trick, too, without the muscle and organ damage or raised risks of heart attacks. Nature always has better options, but nothing beats eating right and exercise, and just a 10 percent loss of body weight can lower those levels quite a bit.

But because the makers of statins know they can rake in billions in profits by scaring the public into thinking they must have the drug, many people still take them without ever questioning their doctors about the possible dangers that may outweigh the benefits.

The Most Dangerous Drugs

According to an August 18, 2019, article by Jena Hilliard for the Addiction Center, these are the most dangerous drugs in terms of overdosing and overuse. In 2019, drug overdoses were the leading cause of accidental death in the United States. The researchers who composed this list tracked death rates and looked at side effects:

1. Acetaminophen (Tylenol™): The most dangerous of the list because of its toxicity to the liver and its widespread use. Related overdoses topped about 50,000 ER visits and 25,000 hospitalizations annually.

2. Alcohol: Second because of the health issues and injuries.

3. Benzodiazepines: Tranquilizers that include Valium and Xanax and accounted for over 31 percent of overdose deaths in 2017 alone.

4. Anticoagulants: Blood thinners such as Warfarin and Xarelto that can cause strokes, heart attacks, pulmonary embolisms, and deep vein thrombosis and can be fatal when combined with aspirin or other blood thinners.

5. Antidepressants: Prozac, Zoloft, Cymbalta, and others prescribed for depression, obsessive-compulsive disorder (OCD), and ADHD and come with a 33 percent increased risk of early death and a 14 percent increase in risk of stroke or heart attack.

Other deadly drugs include bromocriptine, which is used to treat Parkinson's and can cause systematic hypotension and somnolence, episodes of sudden sleep onset; clozapine, an antipsychotic med used to treat schizophrenia and can cause cardiovascular effects such as myocarditis and cardiomyopathy; digoxin, used to treat arterial fibrillation and increases the risk of dying by 20 percent, and semisynthetic opioids such as Vicodin and OxyContin, which are incredibly addicting and dangerous.

These and many other medications can become even deadlier when combined with alcohol, OTC meds, and other prescription meds that have similar side effects, doubling the possibility of overdosing and increasing the risk of death.

Pinkwashed!

Getting diagnosed with breast cancer can be a frightening time. Many people champion Breast Cancer Awareness Month as a time to spread the word about the disease and get support, including money for more needed research. But sadly, AstraZeneca, the Big Pharma giant and maker of the blockbuster breast cancer drug Tamoxifen, is the originator of Breast Cancer Awareness Month, in partnership with the American Cancer Society. If this isn't a bizarre conflict of interest, I'm not sure what is.

AstraZeneca makes two of the biggest drugs pushed on breast cancer patients—Tamoxifen and Arimidex. Tamoxifen is classified by the International Agency for Research on Cancer as a known human carcinogen. But you won't hear that in any of the Breast Cancer Awareness Month campaigns that urge women to get mammograms and other tests for early diagnosis. By promoting mammograms, AstraZeneca can then target its treatments to new breast cancer patients and look like they are compassionate and caring to boot.

Toxic Cocktails

The FDA often doesn't scrutinize drugs as well as it should, resulting in widespread recalls of drugs that might contain carcino-

The company that makes the breast cancer drug Tamoxifen is the same one sponsoring the pink ribbon campaign for awareness of the disease. How likely is it that AstraZeneca truly wants to cure breast cancer and lose a steady income?

genic materials. In fact, they check less than 1 percent of imported drugs for impurities, which is frightening when you consider that about 80 percent of all active drug ingredients are manufactured in China and India, where manufacturing plants rarely undergo inspections. Even if a problem is discovered, it might be a year or more before it is taken off the market and out of the way of consumers.

The compound NDMA, a water-soluble chemical that can cause cancer in animals and is classified as a probable carcinogen in humans, has been found in three widely prescribed blood pressure medications (losartan, valsartan, and irbesartan), the diabetes drug metformin, and in two heartburn meds (Axid and Zantac). They were recalled but only because the NDMA was detected in these specific medications. With under 1 percent ever inspected for purity, you can imagine how many meds out there might be tainted with cancer-causing chemicals and ingredients. NDMA can also cause serious liver failure, too. It has even been used to purposely poison people!

Speaking of cancer, one of the main courses of treatment is chemotherapy. But a new study published in *Nature Cell Biology* in 2019 by an international team of scientists confirmed that chemo spreads cancer. The study examined two chemo drugs that are commonly used, paclitaxel and doxorubicin. The researchers looked at how breast cancer cells responded to these toxins and concluded that they trigger the onset of new tumors in other parts of the body, and they also stated that upon reaching the lungs, they released proteins that resulted in the development of metastatic cancer cells.

 Why on earth would a treatment that is supposed to save lives cause such destruction, even death? Maybe because the global oncology market makes billions of dollars....

Why on earth would a treatment that is supposed to save lives cause such destruction, even death? Maybe because the global oncology market makes billions of dollars and is reportedly on track to make about $200 billion a year by 2022. However, chemotherapy has always been scrutinized as a highly toxic way to fight cancer, the first treatments having been derived from mustard gas, a chemical warfare agent. Some cancer patients today still receive chemo based on these derivatives.

Sadly, as more people are diagnosed with cancer, the pharmaceutical machine makes bigger and bigger profits, and the rates of cancers don't seem to be decreasing. In the book *Victory Over Cancer!,* authors Dr. Matthias Rath and Dr. Aleksandra Niedzwiecki refer to the ninth edition of the book, which states that almost half of the substances listed as causes of cancers are pharmaceuticals. Disease is big business, and cancer is one of the biggest profit-making diseases out there.

Again, there are instances, of course, where pharmaceuticals can save lives or at least extend them, but it behooves the consumer—the sick patient—to ask questions and do research on their own. All these pills and treatments go into your body and your bloodstream, and nobody is as good of an advocate for your health as you will ever be. It may be that the best way to treat what ails you is a combination of methods.

Big Pharma looks at everyone as a chance to make a patient for life. They are in the business not to cure diseases but rather to cover symptoms. If they cure diseases, they are basically all out of a job. The rise in prescribing psychotropic drugs to seniors, children, and veterans is one example of how dangerous treatments can be when you don't ask questions. Some of these antidepressants, antipsychotics, and antianxiety medications can cause more mental health issues than already existed, raise suicide risks, cause paranoia and hallucinations, destroy emotional well-being, and basically add insult to injury by increasing mood, thought, and behavioral disorders.

Yet if you say you are feeling down, or perhaps are a veteran who is suffering from post-traumatic stress disorder (PTSD), or a child who is acting out, one of the first lines of treatment might be a visit to a psychiatrist and a prescription or two for medications that should no longer exist, thanks to decades of treatment, right? Those pills can cause sleep disorders, impaired functioning, high blood pressure, lack of coordination, low sex drive, anger, and violent behavior. Over time, many people become dependent on the drugs and never treat the causes. If they stop working, another drug is added on top of the initial drug. If the patient wants to get off the drugs, often the withdrawal effects are so horrendous, they gladly rush back on and become lifelong patients.

Save Your Cells!

Toxic substances such as mold and chemicals have a huge impact on our cell membrane, which can create oxidative stress and lead to cancers and other diseases. Mycotoxins, heavy metals like lead, chemicals, and viral inflammatory byproducts of meds we take can all disrupt the health of our cell membrane and how it is supposed to function.

 Mycotoxins, heavy metals like lead, chemicals, and viral inflammatory byproducts of meds we take can all disrupt the health of our cell membrane and how it is supposed to function.

The cell membrane is made of a supple phospholipid bilayer and is actively regulating bodily functions associated with energy, fluid regulation, and brain and nerve function. When our cells are exposed to things like mold or toxins, there is a disruption of the electrical communication and signaling function that can then lead to Parkinson's, amyotrophic lateral sclerosis (ALS), cancer, and Alzheimer's, among other diseases. General symptoms of toxic overload can include brain fog, headaches, lethargy, exhaustion, and confusion. By damaging our cell membrane, it sets in motion a domino effect that can affect our overall health and not in positive ways, requiring a detox or reduction of the toxic load in the body.

Toxic ... *Vitamins*???

To improve health, we spend billions of dollars a year on vitamins and supplements, assuming they are free from the toxins and chemicals of Big Pharma medications. Not so fast. Many supplements contain fillers and chemicals, sometimes under names that sound almost natural, that can harm health.

These ingredients, which are on the labels, include inorganic materials such as iron fillings found in cheap vitamins and calcium made from ground-up seashells. The magnesium found in many vitamin/mineral combos is the cheaper magnesium oxide, which does no good to the human body. Some vitamins and supplements come from China and may include lead and arsenic, as well as aluminum, which is also found in many detox products. There are detox liquids with high levels of aluminum and other heavy metals that can cause more harm to the body than the existing toxins they are trying to clear out.

Mineral supplements can also include barium but at low enough levels that they may not be reported.

Vitamins and mineral supplements can cause allergic reactions and side effects if they contain fillers, the sources are cheap, or the dosages are not accurate, even if the label reads "recommended adult dosage." Vitamin A can cause vision changes, hair loss, and skin issues. Vitamin E can lead to vomiting, nausea, headache, fatigue, and blurred vision. Niacin can cause gastrointestinal distress, stomach upset, and flushed skin as well as interact with statins. Not all these side effects will occur in everyone or be dangerous, but it helps to avoid the cheaper products. Do some research into the manufacturer and how they process their products, and watch out for unnecessary fillers.

Vitamins and dietary supplements may contain more in them than you bargain for. Check out the other filler ingredients and talk to your doctor about what you should and should not take.

Dietary supplements also include artificial synthetic food colors, hydrogenated oils added as fillers, the dreaded mercury (especially found in fish oils and fish oil products), and titanium dioxide, which is used as a colorant and can cause serious lung damage, kidney damage, and damage to the DNA when in the form of titanium dioxide nanoparticles.

The labels of many vitamin and mineral supplements claim they are lab tested and approved, but that claim can itself be general. Which lab? Where? The European Union has strict labeling standards for vitamins and minerals, but in the United States and Canada, those strict standards have not yet been fully adopted. For those people who think taking more vitamins is better and will improve their overall health, the truth is high doses can pose serious risks.

Finding Fault

You may blame Big Pharma and the medical community for the overabundance and overprescribing of pills rather than focusing on real health and natural alternatives, but the truth is, there is a huge market they cater to. That market is made up of people who would much rather pop a few pills than take the time and effort to work out and clean up their diets or eliminate stressors. Taking pills has become all too easy, even if it takes years off of our lives with them, but without a market, the pharmaceutical companies would dry up. Instead, they are making more money every year.

Medical schools don't teach doctors how to treat patients with nutrition and exercise regimens, so asking a doctor for help will usually result in a prescription. Fortunately, though, there is a growing body of doctors, researchers, and other medical personnel who are pushing back against Big Pharma and the focus on allopathic medicine and instead partnering with their patients rather than treating them like numbers.

 Medical schools don't teach doctors how to treat patients with nutrition and exercise regimens, so asking a doctor for help will usually result in a prescription.

The patient must learn to stop depending on their doctors to hand them a prescription at every visit like kids who get a lollipop after seeing the dentist. Finding an integrative medicine doctor who tells you to go home and stop eating processed junk foods, take up walking each day, and use herbs for ailments is priceless. Traditional and ancient medicine practices such as Chinese medicine and India's Ayurvedic medicine can treat and manage chronic illnesses effectively. Even "manual medicines" such as chiropractic and osteopathic modalities are rising in popularity because they are more integrative and work.

A few pills may be the easier path to follow, especially for our busy, stressful schedules that don't leave much time for health. But as the sick state of the populace shows, it's not a path worth taking in the long run. Everything you put in your body counts.

What we ingest in the form of toxic food, water, pills, and medications is one thing. What we inject is another beast entirely.

HIDDEN AGENDAS AND CONSPIRACIES

The health and well-being of the nation cannot withstand the massive bombardment of toxins from every direction and thrive. The rise in both major diseases and chronic illnesses has led many to suggest something far more sinister is at work than corporate greed and the lust for profits that has led the pharmaceutical industry to become the most powerful and influential lobby force in the halls of Congress. Add to that the push for more genetically modified organism (GMO) foods, the chemtrails fouling our air, 5G affecting our brains and bodies via electromagnetic (EM) fields, pandemics that don't seem quite natural in origin, and fluoride in almost every municipal water source in the country, and it isn't hard to understand why people might be turning to conspiracies involving the culling of the human population, eugenics, chips in vaccines, and a coming New World Order spoken of for decades, with a tiny but powerful, elite group of politicians, corporate CEOs, and billionaires deciding the fate of humanity while preserving diminishing resources for themselves.

Sounds like crazy talk, but there is plenty of "circumstantial evidence" that these ideas have been around, and in play, for a long, long time. Is the use of toxins in every aspect of our lives a tool or weapon of a greater agenda to keep people sick, weak, and submissive?

So many people have been "red-pilled" in the last few years, wondering if it is all tied together. "Red-pilled" means being awak-

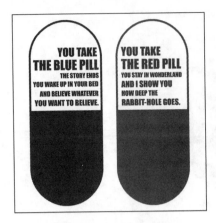

Have you taken the red pill and awakened to what is going on in the environment and products that fill your life?

ened from the "matrix" and is a reference to the 1999 movie *The Matrix* in which the character Neo is told he can take the blue pill, where he will remain blissfully ignorant to the truth, or the red pill, and he will awaken to the truth and go down the proverbial rabbit hole. Those who have been awakened are asking questions, making connections and links between events and situations, and putting together the pieces to a darker aspect of reality that might include people with darker agendas and the power and influence to carry them out. What better way than to use toxic ingredients and chemicals to make people sicker and less likely to fight back, or is it just blatant greed and lack of empathy for humanity driving this alleged quest to poison Americans? After all, there is no equal push to make people healthy using natural things like diet, exercise, and meditation. The focus seems to be on keeping as many people as possible on as many pills and other medications as possible for the longest time possible.

Conspiracy and paranoia or a real rabbit hole worth going down?

Georgia Guidestones and the New World Order

In Elbert County, Georgia, on a high hilltop, there is a huge, granite monument known as the Georgia Guidestones. It is engraved with messages to humanity across the four giant stones that support a common capstone. These ten messages, or guides, are written in eight different languages, one language on each face of the four large, upright stones. Moving clockwise around the structure are English, Spanish, Swahili, Hindi, Hebrew, Arabic, traditional Chinese, and Russian. The messages are:

▲ Unite humanity with a living new language.

▲ Rule passion—faith—tradition—and all things with tempered reason.

▲ Protect people and nations with fair laws and just courts.

▲ Let all nations rule internally resolving external disputes in a world court.

▲ Avoid petty laws and useless officials.

▲ Balance personal rights with social duties.

▲ Prize truth—beauty—love—seeking harmony with the infinite.

▲ Be not a cancer on the Earth—Leave room for nature.

▲ Maintain humanity under 500,000,000 in perpetual balance with nature.

▲ Guide reproduction wisely—improving fitness and diversity.

Just to the west of the monument is a granite ledger level with the ground that identifies the structure and lists facts about the

The Georgia Guidestones are located in Elbert County, Georgia, about 45 miles from Athens. Sometimes called the American Stonehenge (although it was constructed in 1980, which is far from ancient), the stones stand about 19 feet, 3 inches (5.87 meters) tall, and each one weighs 237,746 pounds (107,840 kilograms).

size, weight, and astronomical features of the stones, the date it was installed, and the sponsors. At the top center of the tablet is written:

The Georgia Guidestones

Center cluster erected March 22, 1980

Immediately below this is the outline of a square, inside which is written:

Let these be guidestones to an Age of Reason

Around the edges of the square are translations in four ancient languages. Starting from the top clockwise: Babylonian (cuneiform script), Classical Greek, Sanskrit, and Ancient Egyptian (hieroglyphs).

Known as the "American Stonehenge," the Guidestones were the work of R. C. Christian, although most believe that is not his real name. He walked into the office of the Elberton Granite Finishing Company in June 1979 and said he wanted to build a monument that would act as a message to mankind. He claimed he represented a group of men who sought to offer wisdom and direction for the whole of humanity. No one to date knows who that group is. Some claim he was a part of a group like the Illuminati, who sought a global, one-world order or, as it is better known, the New World Order.

The idea of a New World Order is not a new one. Many will recall President George H. W. Bush speaking of it during his one term in office. It has been the stuff of many books, novels, and movies, straight out of George Orwell's *1984*, a global and unified dictatorship where the powerful people at the very top make all the rules for the dronelike and subservient masses who exist below. The idea of this one world order includes a global government, and a global financing system, under the guise of unifying nations and ridding the world of borders. This secretly rising totalitarian, global government has been considered nothing but conspiracy theory; yet, where there is smoke, there is often fire.

One of the key aspects of the Guidestones was the message referencing a massive decrease in the human population, taking it down to a sustainable 500,000,000. To do this would require the deaths of billions, which some suggest is the goal of the toxic assault on our planet and its peoples. Using everything from chemtrails, 5G, pandemics, poisons in our food and water, vaccines meant to sterilize and kill, and generally keeping most of the world's people sick, tired, and poor seems like a slow way to achieve such a goal, but it isn't entirely all in the realm of conspiracy. Population culling is a concept that has been around for decades, with some powerful names and organizations behind the cause.

No matter who created the Guidestones or for what purpose, they do mirror a larger agenda by the United Nations to achieve similar goals, an agenda that sounds great on paper as a long-term plan to preserve the planet and its precious resources and make the lives of all of humanity better but has raised many conspiracy-minded eyebrows, suggesting a global quest to put into place a number of policies and plans to create a totalitarian New World Order and exterminate a large part of the population.

Depopulation and Eugenics

As the population of Earth climbs, conversations about finding a way to save precious resources rises, too. The idea of depopulation is not new, but as the world grows closer to a globalized new order, the arguments in favor have ramped up in the halls of the elite and powerful, the ones who most stand to gain when the herd is culled and there are more resources to go around for those

This memorial flame was erected in Yerevan, Armenia, to honor the one million Armenians who lost their lives in the genocide by the Ottoman Empire. Genocides have occurred repeatedly in human history, including the Jewish Holocaust and several ongoing genocides

at the top. Toxins might be one of the weapons of depopulation, a slow death of millions from medications, food, air, and water, coupled with more global pandemics, lockdowns, chemtrail activity, and a blanket of 5G to put the final fix in for the 99 percent struggling to survive and have decent lives at the bottom of the lopsided pyramid of power.

During the Holocaust, Germans talked of the disabled, mentally ill, elderly, and terminally ill as *nutzlose Fresser*, which translates to "useless eaters." Genocidal depopulation was the goal of the Final Solution, and today, some claim it is the goal of the chemical assaults on our bodies, meant to sicken and weaken, sterilize, and even kill as many useless eaters as possible. With the global population at 7.8 billion and expected to hit 9 billion in a few years, wouldn't it be in the best interest of the planet to cull the herd enough to sustain a certain way of life, just as wild herds of animals are culled by hunting and disease? It might make sense to the elite, who know they won't be culled, but to those who don't win the lottery of life, it is a terrifying, brutal, and inhumane way of dealing with the potential problem of overpopulation. It is nothing short of playing God by a preselected group of people who decided they were gods without public input, notably the richest families on Earth. The Rothschilds. Rockefellers. The Gates and Soros families. The British Royalty and American political "royalty," and, of course, the heads of corporations, banks, and other institutions of power and influence. Some would include the Illuminati, the Bilderberg group, the Trilateral Commission, Freemasons, and the "Deep State," shady and shadowy entities that are the man behind the curtain in the *Wizard of Oz* story, pulling the levers and strings.

 It is nothing short of playing God by a preselected group of people who decided they were gods without public input, notably the richest families on Earth.

Yet those arguing against a depopulation agenda claim that the elite wouldn't kill off their workforce—the slaves and drones at the bottom of the ladder rungs that do all the work that make them millions. Heavy depopulation would be an economic disaster and potentially bring about a collapse of government, not to mention a collapse of all services and production. If corporations and Wall Street are a part of the culling of the useless eaters, wouldn't that leave fewer useless eaters to buy and consume the products they are making, selling, and trading on?

Recall China's former "one child policy" that limited families to having one child as a means of keeping the population sustainable. After a baby boom during the 1960s, the Chinese government realized it had too many mouths to feed as a result and then sought to undo the damage by urging citizens to only have one child, which was met with negative reactions from the public and other world governments. The birth rate ended up plummeting so low, there was not a stable workforce left and an off-balance ratio of males to females since most families favored boys and aborted girl babies. Ultimately, the plan was revised to allow for two births.

That wasn't the first time a low birth rate backfired. During Joseph Stalin's reign in Russia, he taxed childless people to encourage them to procreate and gave out medals and rewards to women who had many babies. The result was a boost in population that went too far in the other direction and caused the eventual end of the reward system! Nazi Germany banned abortions, contraception, and homosexuality as a way of boosting its low birth rate.

Plagues and pandemics worked wonders in the past to lower the population, especially of the already sick and ailing. From 1346 to 1353, the Black Death of Europe reduced populations there by

The Black Death that struck Europe in the Middle Ages killed half the population. Plagues are an effective way to cull a population. Some conspiracy theorists think that a plot is afoot to do just that.

TOXIN NATION

between 75,000,000 and 200,000,000. During the Spanish Flu of 1918, the Rockefeller Foundation, considered the founder of the "pharmaceutical cartel," at the forefront of a new global business venture called the pharmaceutical investment business, actively started a witch hunt, accompanied by a complicit media, against any medicines or treatments for the plague it did not own the patent for or control. No matter how many lives were lost, the Rockefeller machine, the most powerful in the land at the time, suppressed anything, including a rising vitamin industry, that wasn't made or manufactured in its own labs.

The emergence of AIDS and HIV in the 1980s was thought by many to be a new culling of the herd that focused on gay men and eventually made its way into the general population. Could today's COVID-19 pandemic be a population-control agenda in play? In a chilling interview for *People* magazine in 1981, England's Prince Philip, Duke of Edinburgh, who was known to be a globalist, stated that human population growth was the single most serious long-term threat to survival and that if it wasn't soon curbed, we would be in for a major disaster. He claimed the rising population would be competing for resources, creating more pollution. He said that if it wasn't controlled voluntarily, it

would be controlled "involuntarily by an increase in disease, starvation, and war." He was quoted later as stating, "In the event I am reincarnated, I would like to return as a deadly virus, in order to contribute something to solve overpopulation."

In 2012, media mogul, billionaire, and population-control advocate Ted Turner told journalists he would like to reduce the global population by five billion people, urging a one-child policy for the world for 100 years. He founded CNN, the major news outlet that has served as a liberal voice in the media for decades. Turner claimed population control was a longtime passion and goal, so long as it was on a voluntary basis. But this was several years ago, and with the population fast approaching the eight-billion threshold, he might have changed that tune. His donation to the UN in 1997 of $1 billion

Media mogul Ted Turner once declared that we should reduce the world's population by five billion people.

to tackle "women and population issues" certainly speaks of an ongoing desire to fulfill his desires.

It certainly wouldn't be the first time billionaires expressed a desire to control or lower the population, and it wouldn't be the last. In 2009, a group of billionaires met in New York, including Turner, Bill Gates, Warren Buffett, David Rockefeller, Oprah Winfrey, Michael Bloomberg, and George Soros, to discuss trying to cut back the world's population. They called this friendly little get-together "The Great and the Good," but it seemed more like the "rich and the elite" gathering to drink and eat and discuss ways to promote charitable giving while also deciding on potential plans to cull humanity down to their objectives. One must imagine how these kinds of meetings have played into the "humanitarian" agendas we face today.

Forced Sterilizations

Women in the United States have been subjected to forced sterilization and other coordinated efforts to control their fertility, especially after the women's rights movement emerged out of the 1960s. According to "The History of Forced Sterilization in the United States" by Kathryn Krase for the *Our Bodies, Ourselves* website, about one-third of all Puerto Rican women ages twenty to forty-nine were sterilized as a part of a population-control effort that was designed by a eugenics board that chose sterilization as a means of keeping the population down. This was most often done by tubal ligation and hysterectomy, all without any kind of informed consent.

Even women on the U.S. mainland were subject to sterilization policies that had been enacted as early as 1907, giving the government the right to "sterilize unwilling and unwitting people." Thirty states passed such laws and targeted the mentally ill, disabled, diseased, and dependent, stating they were unable to regulate their own reproductive abilities. Often, women were chosen by racial and class groups. Eugenics boards were set up in the early twentieth century to decide who got sterilized, and this often included men. According to the article, between 1930 and 1970, North Carolina sterilized over 7,600 people.

Ethnicity and race often played into who was most targeted for sterilization, with Latina women specifically targeted in the twentieth century in Los Angeles, New York, and Puerto Rico. Black women were also being sterilized in far greater numbers than white women, especially in places like North Carolina and others that were noted for "discriminatory sterilization." Native American women were also subjected to "coercive population control" throughout the twentieth century. These were all a form of ethnic cleansing and fell under the category of eugenics. But before we go there, in 2013, a report found that almost 150 women were illegally sterilized in California prisons between 2006 and 2010, using state funds because federal funds did not allow sterilization of incarcerated women.

Though tubal ligation and hysterectomies have often been used in sterilization, these are complicated procedures. Why not just inject them with a vaccine? Vaccines to control fertility in both men and women are under development at the National Institute of Immunology in New Delhi, India, where a single injection procedure that can castrate or sterilize a male depending on the injection site has passed through field tests and should be on the market soon. Another vaccine that induces antibodies against the human chorionic gonadotropin hormone has been passed through the testing phase. Other types of vaccines targeting other avenues of sterilization are undergoing testing procedures.

 It is one thing to knowingly take a vaccine to render yourself sterile but quite another to have one forced on you under the guise of a government-mandated population-control effort.

Perhaps this will be the new choice of birth control for many men and women, but the potential for abuse is huge. It is one thing to knowingly take a vaccine to render yourself sterile but quite another to have one forced on you under the guise of a government-mandated population-control effort. Worse still, how would you know which vaccine you were getting when you went to your doctor for a flu shot? If those who believe in population control have any say, you won't.

Just ask the women of Kenya who were given vaccines that made them sterile against their knowledge. This happened in the early 1990s as part of a UN plan to carry out a tetanus vaccination campaign among young women in developing countries without telling these women they were being vaccinated against pregnancy with the same shot. This also happened in Mexico in 1993, the Philippines and Nicaragua in 1995, and Peru in 1995.

In 2014, in Kenya, the Kenya Catholic Doctors Association and the Kenya Conference of Catholic Bishops called out the UN agencies when they attempted a second vaccine program on young women. All twenty-seven Kenyan bishops signed a statement concerning the World Health Organization (WHO)/UNICEF tetanus vaccination campaign, stating they believed it was part of a larger population-control agenda.

Eugenics

The desire to lower the human population often plays into a belief that getting rid of the weakest, poorest, and often the minorities will strengthen, purify, and improve the genetic quality of humanity in the long run. This is called "eugenics" and goes all the way back to Plato advocating selective breeding in 400 B.C.E. The goal was to improve the quality or stock of human beings, so they would be the most superior of the bunch, getting rid of the inferior members of the species via a variety of methods.

White supremacists, Nazis, and bioethicists have used modern eugenics to "improve" the human race by promoting specific traits or characteristics. Sometimes that includes race. Modern eugenics began in the early part of the twentieth century with programs in the United States, Canada, and Europe designed to improve the human population and discourage reproduction among those considered unfit. This could be achieved via prohibiting marriage and childbirth, sterilization, and incarceration.

The most blatant example is the extermination of Romanis and Jews in Nazi Germany. Adolf Hitler's minions believed that by ridding the world of these groups, they would purify the human race through their horrific abuses and the Final Solution. The desire to genetically alter humans for their betterment was one that was carried on poor victims during the Holocaust in the deadly labs of such Nazi experimenters as Josef Mengele.

Positive eugenics strives to better the species. In ancient times, children were inspected by town elders. Those deemed unfit were killed—often drowned—especially the deformed or sick. In more modern times, such figures as Winston Churchill and the powerful Rockefeller family supported

The late nurse, sex educator, and birth control advocate Margaret Sanger was widely criticized for supporting eugenics.

eugenics. Current names in the news associated with eugenics beliefs include Microsoft cofounder Bill Gates and the late birth control advocate Margaret Sanger.

Sanger, the founder of Planned Parenthood (PP), was recently vilified for her eugenics and her focus on targeting blacks for abortions. She was such a racist and eugenicist that Planned Parenthood of New York removed her name from its Manhattan Health Center, a decision followed by more PP facilities that didn't want to be associated with a woman who championed birth control and supported racist ideas that included lowering the population of blacks, whom she deemed as genetically inferior. PP has been providing services for women, including healthy checkups and breast cancer screenings, for years, but their abortion history is a dark mark that may not vanish because their founder's name was removed from a few buildings.

Gates is one of the most dominant figures in modern history despite coming from a technology background. Today, he decides policies involving vaccines, pharmaceuticals, genetically modified organisms (GMOs), food production, and more, yet he has no training in any of these fields. Money apparently buys power and influence, but his family set the stage for his own belief in lowering the population, and he's on video stating one of the ways to do this is with vaccines.

Computer company mogul Bill Gates amassed a fortune that now makes him highly influential in the world of politics and business, including the medical world.

His father, Bill Sr., and his mother, Mary, amassed great wealth and influence, and both were proponents of eugenics, passing on that legacy to their son. Gates Sr. once ran Planned Parenthood, as revealed by his son in a 2003 interview with Bill Moyers. He recounted his own passion for funding reproductive issues because of his father. In the same interview, Gates claimed his desire to reduce the human population was not about getting rid of the useless eaters but rather a true desire to improve health and education for the people on the planet. By doing so, it would decrease the number of children those people would have.

His mother was the president of King County, Washington's United Way and in 1983 became the first woman to be appointed the national group's executive committee chair. As Bill Gates became more powerful and influential, he managed to enter the greatest halls of power with his massively funded Bill and Melinda Gates Foundation, which is now linked to dozens of pharmaceutical companies, vaccine companies, research institutions, politicians, and businesses. Gates is concerned with global vaccines that include chips as digital IDs, making humans walking digital credit cards and information sources to be tracked, traced, and controlled by any outside entity with tech know-how. And the way to introduce those chips? Vaccines. Because of his funding of large organizations that dictate global food policy, you could say he not only controls what we ingest and consume but also what we have injected into us.

Agenda 2030

The United Nations' Agenda 21 has been cited by conspiracy researchers and theorists as a plan to cull the human population to bring about a more sustainable planet, suggesting that the limited resources of land, water, and food will be kept for those deemed worthy of living. Under the guise of a "plan for sustainability," skeptical eyes see the drive toward a one-world system where no one will own anything and only the strongest will survive.

The United Nations (General Assembly in New York City pictured) wrote what it calls Sustainable Development Goals (also called the 2030 Agenda or Agenda 2030), an outline of 17 goals with the idea of ending poverty, hunger, inequality, and unsustainable development.

The text of Agenda 21, and its revised version, Agenda 2030, are below, although not in their entirety. It's hard to argue with the belief we need to do more to protect our planet and its precious resources, but talk of population and the emphasis on globalization are what create a shadow over these lofty goals.

Agenda 21

Agenda 21 is a comprehensive plan of action to be taken globally, nationally and locally by organizations of the United Nations System, Governments, and Major Groups in every area in which human impacts on the environment.

Agenda 21, the Rio Declaration on Environment and Development, and the Statement of principles for the Sustainable Management of Forests were adopted by more than 178 Governments at the United Nations Conference on Environment and Development (UNCED) held in Rio de Janeiro, Brazil, 3 to 14 June 1992.

The Commission on Sustainable Development (CSD) was created in December 1992 to ensure effective follow-up of UNCED, to monitor and report on implementation of the agreements at the local, national, regional and international levels. It was agreed that a five year review of Earth Summit progress would be made in 1997 by the United Nations General Assembly meeting in special session.

The full implementation of Agenda 21, the Programme for Further Implementation of Agenda 21 and the Commitments to the Rio principles, were strongly reaffirmed at the World Summit on Sustainable Development (WSSD) held in Johannesburg, South Africa from 26 August to 4 September 2002.

As the time got closer to 2021, it was deemed necessary to revise the Agenda, which is now called Agenda 2030. To those who oppose it, it still smacked of a globalized world under a new world order where individual rights are dismantled, resources are protected at all costs, and the population is lowered to make the world more sustainable for those lucky enough to remain, most likely the elite and the wealthy. Are conspiracy theorists reading between the lines, or is this Agenda nothing more than the nations of the planet coming together to make the world a better place?

The Revised Agenda 2030 Preamble

Transforming our world: the 2030 Agenda for Sustainable Development

Preamble

This Agenda is a plan of action for people, planet and prosperity. It also seeks to strengthen universal peace in larger freedom. We recognise that eradicating poverty in all its forms and dimensions, including extreme poverty, is the greatest global challenge and an indispensable requirement for sustainable development. All countries and all stakeholders, acting in collaborative partnership, will implement this plan. We are resolved to free the human race from the tyranny of poverty and want and to heal and secure our planet. We are determined to take the bold and transformative steps which are urgently needed to shift the world onto a sustainable and resilient path. As we embark on this collective journey, we pledge that no one will be left behind. The 17 Sustainable Development Goals and 169 targets which we are announcing today demonstrate the scale and ambition of this new universal Agenda. They seek to build on the Millennium Development Goals and complete what these did not achieve. They seek to realize the human rights of all and to achieve gender equality and the empowerment of all women and girls. They are integrated and indivisible and balance the three dimensions of sustainable development: the economic, social and environmental.

The Goals and targets will stimulate action over the next fifteen years in areas of critical importance for humanity and the planet:

People

We are determined to end poverty and hunger, in all their forms and dimensions, and to ensure that all human beings can fulfil their potential in dignity and equality and in a healthy environment.

Planet

We are determined to protect the planet from degradation, including through sustainable consumption and production, sustainably managing its natural resources and taking urgent action on climate change, so that it can support the needs of the present and future generations.

Prosperity

We are determined to ensure that all human beings can enjoy prosperous and fulfilling lives and that economic, social and technological progress occurs in harmony with nature.

Peace

We are determined to foster peaceful, just and inclusive societies which are free from fear and violence. There can be no sustainable development without peace and no peace without sustainable development.

Partnership

We are determined to mobilize the means required to implement this Agenda through a revitalised Global Partnership for Sustainable Development, based on a spirit of strengthened global solidarity, focused in particular on the needs of the poorest and most vulnerable and with the participation of all countries, all stakeholders and all people.

The interlinkages and integrated nature of the Sustainable Development Goals are of crucial importance in ensuring that the purpose of the new Agenda is realised. If we realize our ambitions across the full extent of the Agenda, the lives of all will be profoundly improved and our world will be transformed for the better.

CONCLUSION: KILLING OURSELVES TO LIVE?

As life gets faster, cheaper, and easier, human health has taken a back seat to convenience and accessibility. What we can easily obtain, we think we most need, and maybe that's true for some things, but certainly not for our well-being. Our health is tied directly to the habits we form, the products we buy, and how we choose to nourish our bodies, but all too often, we take the easy road and, if we have a headache, pop a pill. If we feel tired, we pop a pill or eat junk food. If we are stressed, we pop a pill, eat junk food, and sit on the couch with a big glass of wine. Toxicity becomes habitual, only because it's usually the easy way out. We know toxins harm us, but we don't seem to stop them from being a huge part of our lives. We know that the research backs up the disturbance of our immune systems and brains when exposed to high rates of electromagnetic fields (EMF), but we cannot wait for our new 5G phones.

It also becomes habitual when we remain ignorant to it and don't take the time necessary to look at the labels, ask about the ingredients, and research how things are made, and with what, all of which influence our health. When examining the toxins we absorb and ingest, breathe in and inject, consume and expose ourselves to become less important than what's on Netflix to binge watch or what is happening on our Instagram page, we have a problem. Not only do we leave ourselves open and vulnerable to a host of adverse side effects, even death, but we also tell the companies selling us the pills, vaccines, and toxins that we are submissive and will take whatever they give us, no resistance and no questions asked.

 Whether or not we believe in conspiracy theories ... we do know one thing is true: we are being poisoned in huge and unprecedented ways.

Whether or not we believe in conspiracy theories about a global system working against us, trying to deprive us as individuals of our basic right to health and happiness, we do know one thing is true: we are being poisoned in huge and unprecedented ways. Our air, water, food, products, medications, vaccines ... the list keeps getting longer, and the methods of poisoning are more invasive and pervasive than ever. Perhaps there is a greater sinister agenda behind this. Perhaps it is just the fallout of massive corporate greed and the complicit people getting richer and more powerful because of it. It really may not matter because the final truth is, we can stop the toxic shock—the poisoning—by standing up to it first in our own bodies, our own families, our own homes, our own communities, and our own governments. Yet why haven't we stopped the bleeding? Could it be fear? Are we allowing our rights to be stripped away on individual and collective levels by a global body of corporations and people who want nothing more than totalitarian control over us? Is removing our governance and sovereignty over our own bodies and the choices we make the first step to turning our entire lives over to these mysterious others?

We turn our power over all the time. We did so after 9/11, and we will again when the next great challenge is upon us.

It would be easy and understandable to look at the many toxins we are exposed to on a day-to-day basis and feel overwhelmed and disempowered to change anything. But we can change things. Laws are being written based upon the demands of citizens and consumers who have had enough of the poison in their air, water, food, and products. People are taking on their local political officials, as well as those on a national level. The pressure must be steady, though, because the powerful lobby groups that produce the toxins are always watching, always waiting to squash any citizen uprising and demand.

So, what can we do? We can stop killing ourselves to live, so to speak, and start really living healthier lives by getting rid of toxins in our homes and workspaces. It begins with our bodies and

By making wiser choices in what we purchase and consume, we can not only reduce the toxins within ourselves but can also stop encouraging corporations from using poisons.

detoxing our livers and organs from the many chemicals we have ingested, inhaled, and consumed. We can then look at ways to clean up our homes and buy products that are eco-friendlier and less likely to cause us physical harm. We can demand that our workplaces do the same. If we use our pocketbooks and wallets, eventually those at the top will notice and think twice about continuing to poison everything we consume and use. Speaking vocally and speaking with the money we spend are both powerful ways to fight back and reclaim our health.

Too many people want cheaper, faster, easier, and that is how we end up killing ourselves in our quest to live better, and we often don't even think about it. We eat and buy products that have chemicals in them, pop pills that damage our health beyond the reason we popped the pill, inject our bloodstreams with toxins, and fill our homes with products that emit gasses, vapors, dust, radon, and lead, and we think we are improving our lives and lifestyles. It's a huge wake-up call to realize we do have control over how many toxins we live with. What we can control, we should, even if it costs a little more money or is a little harder to come by.

We can start shopping locally, buying foods at farmer's markets or local farms and ranches, and purchasing products that state they do not contain the toxins we want to avoid. We can buy hand-made gifts and patronize mom-and-pop eateries and stores where we are more likely to get whole foods and real products that were made with far more love and far fewer chemicals. We can also learn to make our own cleaners, soaps, and other items instead of going the cheap-and-easy route and buying them in a store or online marketplace. They don't need our money, and they don't care what they sell us. But we should care.

When it comes to getting rid of toxic substances in our homes, we can contact our local recycle center or city hall to get proper disposal instructions. This applies to paints and solvents, motor oil, pool chemicals, auto fluids, cleaners, fertilizers, batteries, LED light bulbs, computers and electronics, old television sets, and anything else you know shouldn't be tossed into the recycle can or bin (which often list what is or is not acceptable on the underside of the lid). Calling your trash company might do the trick, as they often have hazardous waste pickups and drop-offs. Local schools and businesses will post signs when they hold electronic waste or old mattress collections. Often, these items can be recycled or used for parts, so please don't stick them in the regular trash, where they end up in landfills for years and years to come.

 There is no excuse but laziness and perhaps a lack of knowledge for throwing away such dangerous substances and household hazards in the regular trash....

There is no excuse but laziness and perhaps a lack of knowledge for throwing away such dangerous substances and household hazards in the regular trash, where they end up in landfills and pollute the air and soil for decades, if not longer, and even make their way into local water sources. These toxins then live on to poison our children, and their children, for decades to come.

Learning as much as we can about toxins is easy, too. We can do online research, go to the library, or visit state and local websites for resources on what toxins are in our own water and air, all of which have plenty of information to offer on how to identify toxins in the home and office and what to do about them. Many of these websites are mentioned in this book, including the Environmental Working Group, Children's Health Defense, ICAN, NaturalHealth365, GreenMedInfo, and others, all working to get out information that is often censored by the mainstream media and Big Pharma, Big Agribusiness, and Big Business in general. The Centers for Disease Control and Prevention (CDC), for all its faults and conflicts of interest, does provide a good amount of research and general information about toxins as well as the Food and Drug Administration (FDA) and the Environmental Protection Agency (EPA). Research everything before accepting anything is a great rule of thumb.

There are so many health websites and experts with ideas and methods for detoxing our bodies from the toxins we've ingested. This may be anything from a juice fast to eating more kale, but always check with your doctor before trying any health detox. Web searches pull up tons of articles and research studies on the latest developments with various toxins and poisons in our food and products, but it's important to look for good scientific sources and research. Visit forums where real people talk about drug and vaccine side effects, as you won't hear these stories in mainstream media.

Sometimes, it comes down to just taking a minute to read a label on every product you consider buying. People will spend hours going through memes on social media but not take the same amount of time to research the food and medications, including vaccines, they put into their bodies and the bodies of their children. This "I don't have time" attitude allows us to play right into the

Teach your kids to read labels and make wise choices for what they put in their bodies. It's important to take time, be informed, and teach children good habits.

hands of the companies and corporations that make billions off our laziness and lack of interest as they push more toxic products upon us, telling us why we need them and how great they will make our lives. We buy it all because it is easier to submit and hope for the best than it is to dig into the motives and agendas behind it.

Right now, there is ample censorship and cancel culture of many voices and companies that truly do operate in the best interest of the consumer. But these alternative voices are out there with plenty of science behind them if you take the time to look. Checking product labels is the first step, but it does you no good if you don't know what the ingredients are and how they might harm your body. Research is critical to ultimately detoxing yourself of the foods, water, air, products, and medications that are all around you, to the best of your ability. It isn't an easy process, but it could save your life or, at the very least, add quality years to it.

We don't have to keep killing ourselves to live better. It's a habit we can break and one that will not only better serve us but our children and the planet as well. We may not have choices over every toxin in the environment, but we do have choices over a lot of things, such as where we live, what we put in our bodies, the clothes we wear, the medications we take, and how we furnish our homes.

By taking control of what we can, our health and well-being benefits, and we teach future generations how to live with less toxic exposure so they, too, can experience long, healthy, and happy lives. In the mountains of Northern Pakistan, the Hunza people, a community of 87,000, have the highest longevity in the world, rarely suffer from diseases, and experience great levels of happiness. They have been reported to live up to 120 years—even as long as 160 years—and to retain their looks and beauty in their later years. Women in the Hunza community can have babies even at the age of sixty-five. They eat only healthy foods, twice a day, and do a lot of physical labor. Their diet consists of raw and whole foods like fruits and vegetables, nuts, cheeses, milk, and yogurt, all unprocessed. They eat grains that they grow themselves and drink glacial water. They do not drink alcohol or smoke.

Though you may not want to live to be 160, it is possible when you are not exposed to the level of toxins the average person in the United States must deal with. The human body is powerful and resilient, and when treated right, will return the favor for many years to come.

FURTHER READING

"50 Million Gallons of Toxic Water Flows into Lakes and Streams Every Day in the US." *Associated Press*, February 22, 2019.

"50 U.S. Code § 1520A—Restrictions on Use of Human Subjects for Testing of Chemical or Biological Agents." Legal Information Institute. Accessed July 26, 2021. https://www.law.cornell.edu/us code/text/50/1520a.

"5G Technology and Induction of Coronavirus in Skin Cells." *Journal of Biological Regulators and Homeostatic Agents*, June 9, 2020.

Adl-Tabatabai, Sean. "California Firemen: Wildfires Caused by 'Directed Energy Weapons'—Media Blackout." News Punch, January 10, 2019. https://newspunch.com/california-wildfires-directed-energy-weapons/.

Alton, Lori. "Statin Drugs Offer Serious Health Problems." NaturalHealth365, September 10, 2020. https://www.naturalhealth365.com/statin-drugs-cholesterol-3509.html.

Armstrong, Martin. "'Are They Planning ID2020 as Mandatory Implants for All as the Solution to the Crisis?'" *Armstrong Economics*, March 19, 2020.

"Article36." Accessed July 26, 2021. https://www.article36.org/.

Begich, Nick, and Jean Manning. "Angels Don't Play This HAARP: Advances in Tesla Technology." *Earthpulse*, 1995.

"Bottled Water Found Contaminated." Safe Space Protection, January 5, 2016. https://www.safespaceprotection.com/news-and-info/bottled-water-found-contaminated/.

"Bottled Water vs. Tap Water: Pros and Cons." Medical News Today. MediLexicon International. Accessed July 26, 2021. https://www.medicalnewstoday.com/articles/327395.

Bruek, Hillary. "What Drinking from Plastic Water Bottles Does to Your Body." Business Insider, August 23, 2019. https://www.businessinsider.com/what-drinking-plastic-does-to-your-body-2018-3.

Burks, Fred. "HAARP: Secret Weapon Used for Weather MODIFICATION, Electromagnetic Warfare." Global Research, September 24, 2019. https://www.globalresearch.ca/haarp-secret-weapon-used-for-weather-modification-electromagnetic-warfare/20407.

Cantwell, Allen. "'AIDS and the Doctors of Death: The Evidence of a Secret Genocide Plot.'" New Dawn Magazine, no. 129 (2011).

"Causes, Effects and Wonderful Solutions to Environmental Pollution." Conserve Energy Future, July 2, 2020. https://www.conserve-energy-future.com/causes-and-effects-of-environmental-pollution.php.

Children's Health Defense Team. "Chlorpyrifos: Playing PESTICIDE Politics with Children's Health." Children's Health Defense, October 27, 2020. https://childrenshealthdefense.org/news/chlorpyrifos-playing-pesticide-politics-with-childrens-health/.

Chossudovsky, Michel. "The Ultimate Weapon of Mass Destruction: 'Owning the WEATHER' for Military Use." Global Research, September 15, 2018. https://www.globalresearch.ca/the-ultimate-weapon-of-mass-destruction-owning-the-weather-for-military-use-2/5306386.

Cole, Leonard A. Clouds of Secrecy: The Army's Germ Warfare Tests over Populated Areas. Baltimore, MD: Rowman & Littlefield, 1990.

"Current State of the Ozone Layer." EPA. Environmental Protection Agency. Accessed July 27, 2021. https://www.epa.gov/ozone-layer-protection/current-state-ozone-layer.

Davis, Devra. Disconnect: The Truth about Cell Phone Radiation, What the Industry Has Done to Hide It, and How to Protect Your Family. New York: Dutton, 2010.

"A Deep Dive into the Conspiracy Theory That Governments Are Controlling Us with Fluoride." VICE. Accessed July 26, 2021. https://www.vice.com/sv/article/kwz5m3/why-are-governments-putting-fluoride-in-our-water-sheeple.

Degsy. "The Georgia Guidestones Mystery." The Secret Truth About, May 10, 2020. https://thesecrettruthabout.com/the-georgia-guide stones-mystery/.

DeSaulniers, Veronique. "Microwaves Shown to Cause Serious Harm." NaturalHealth365, September 9, 2020. https://www.natu ralhealth365.com/microwaves-emf-pollution-2829.html.

DiFiore, Tomas. "Chemtrails and CLIMATE Geo-Engineering: The Shasta Decision." Indybay. Accessed July 26, 2021. https://www.in dybay.org/newsitems/2014/08/14/18760026.php.

"Electromagnetic Radiation and Other Weapons of Mass Mutation." Waking Times, May 17, 2017. https://www.wakingtimes.com/ 2015/11/30/electromagnetic-radiation-other-weapons-mass-mu tation/.

"Exposure to Chemicals in Plastic." Breastcancer.org, September 11, 2020. https://www.breastcancer.org/risk/factors/plastic.

"Fluoride's Direct Effects on Brain: Animal Studies." Fluoride Action Network, September 4, 2019. http://fluoridealert.org/studies/ brain04_/.

Freeland, Elana. *Chemtrails, HAARP, and the Full Spectrum Dominance of Planet Earth*. Port Townsend, WA: Feral House, 2014.

———. "The Electromagnetic Space Age We Live In." The Electromagnetic Space Age We Live In. Accessed July 28, 2021. http:// www.positivehealth.com/article/environmental/the-electromag netic-space-age-we-live-in.

———. *Under an Ionized Sky: From Chemtrails to Space Fence Lockdown*. Port Townsend, WA: Feral House, 2018.

Freeman, Makia. "5G Danger: 13 Reasons 5G Wireless Technology Will Be a Catastrophe for Humanity." Global Research, February 2, 2021. https://www.globalresearch.ca/5g-danger-13-reasons-5g-wireless-technology-will-be-a-catastrophe-for-humanity/5680503.

Gandhi, Renu, and Suzanne M. Snedeker. "Consumer Concerns about Hormones in Food," 2000. https://ecommons.cornell.edu/ bitstream/handle/1813/14514/fs37.hormones.pdf.

"GMO Foods: What You Need to Know." *Consumer Reports*, February 2015.

Graham, Ben, and Peter Bodkin. "Thousands of Aussies Engage in 'AGENDA 21' CONSPIRACY." NewsComAu. news.com.au, December 21, 2019. https://www.news.com.au/technology/online/social/tens-of-thousands-of-aussies-engaging-in-agenda-21-conspiracy-on-social-media/news-story/c977779fd74d60ac69e84cb07b56f483.

Group, Edward. "Fluoride: Daily Exposure to Poison." Dr. Group's Healthy Living Articles, November 13, 2015. https://globalheal ing.com/natural-health/fluoride-poison-on-tap-documentary/.

Hackney, Thomas N. "Conspiracy Theory: The Incredible Story of the Cancer Treatment Krebiozen." *Paranoia Magazine*, Fall 2013.

Haslam, Edward T. *Dr. Mary's Monkey: How the Unsolved Murder of a Doctor, a Secret Laboratory in New Orleans, and Cancer-Causing Monkey Viruses Are Linked to Lee Harvey Oswald, the JFK Assassination and Emerging Global Epidemics.* Walterville, OR: Trine Day Publishing, 2007.

Helmenstine, Anne Marie. "Chemtrails versus Contrails." ThoughtCo. Accessed July 26, 2021. https://www.thoughtco.com/chemtrails-versus-contrails-3976090.

Hilliard, Jena. "The Top 15 Most Dangerous Drugs." Addiction Center, November 20, 2020. https://www.addictioncenter.com/news/2019/08/15-most-dangerous-drugs/.

Hood, Ernie. "Are Edcs Blurring Issues of Gender?" Environmental Health Perspectives. National Institute of Environmental Health Sciences, October 2005. https://www.ncbi.nlm.nih.gov/pmc/ar ticles/PMC1281309/.

"Household Air Pollution and Health." World Health Organization. World Health Organization. Accessed July 28, 2021. https://www .who.int/news-room/fact-sheets/detail/household-air-pollution-and-health.

Johnston, Ian. "Male Fish Are Becoming Female Because of Birth Control in the Waters." The Independent, July 3, 2017. https://www.independent.co.uk/environment-fish-changint-sex- gender-chemicals-rivers-water-charles-tyler-fisheries-a7821086 .html.

Jones, Marie D., and Larry Flaxman. *Mind Wars: A History of Mind Control, Surveillance, and Social Engineering by the Government, Media, and Secret Societies.* Pompton Plains, NJ: New Page Books, 2015.

Krase, Kathryn. "The History of Forced Sterilization in the United States." Our Bodies Ourselves, September 21, 2020. https://www.ourbodiesourselves.org/book-excerpts/health-article/forced-sterilization/.

LaVille, Sandra, and Matthew Taylor. "'A Million Bottles a Minute: World's Plastic Binge 'As Dangerous as Climate Change." *The Guardian*, June 28, 2017.

Leitenberg, Milton. "Did the SARS-CoV-2 Virus Arise from a Bat Coronavirus Research Program in a Chinese Laboratory? Very Possibly." *Bulletin of the Atomic Scientists*, June 4, 2020.

Loria, Kevin. "Over and Over Again, the Military Has Conducted Dangerous BIOWARFARE Experiments on Americans." Business Insider, September 25, 2016. https://www.businessinsider.com/military-government-secret-experiments-biological-chemical-weapons-2016-9.

Mackenzie, Jillian. "Air Pollution: Everything You Need to Know." NRDC.org, November 1, 2016. https://www.nrdc.org/air-pollution-everything-you-need-to-know.

Marrs, Jim. *Population Control: How Corporate Owners Are Killing Us*. New York: HarperCollins, 2015.

Marti, James E. *Alternative Health and Medicine Encyclopedia*. Detroit, MI: Visible Ink Press, 1998.

Mary Kekatos Health Reporter for Dailymail.com. "Study Claims Taking Tylenol during Pregnancy Doubles Adhd, Autism Risks for Kids." Daily Mail Online. Associated Newspapers, October 30, 2019. https://www.dailymail.co.uk/health/article-7630583/Study-claims-taking-Tylenol-pregnancy-doubles-ADHD-autism-risks-kids.html.

"Microwave Oven DANGERS." Safe Space Protection, September 23, 2016. https://www.safespaceprotection.com/news-and-info/microwave-oven-dangers/.

Millennium Report. "Directed Energy Weapons Now Being Used Around the World." The Millennium Report, January 1, 1973. https://www.themillenniumreport.com/2018/17/directed-energy-weapons-now-being-used-acround-the-world/.

Paddock, Michael. "What Is Arsenic Poisoning?" *Medical News Today*, January 4, 2018.

Pearce, Fred, and Adam Popescu. "Geoengineer the Planet? More Scientists Now Say It Must Be an Option." Yale E360. Accessed July 27, 2021. https://e360.yale.edu/features/geoengineer-the-planet-more-scientists-now-say-it-must-be-an-option.

Phonegate, Equipe. "Increased Incidence of Brain Cancer. Caused by Exposure to Pesticides and Electromagnetic Fields?" Global Research, December 5, 2019. https://www.globalresearch.ca/brain-cancers-4-times-more-new-cases-glioblastoma-2018-according-public-health-france/5696844.

Piper, Kelsey. "Why Some Labs Work on Making Viruses Deadlier—and Why They Should Stop." *Vox*, May 1, 2020.

Pizzorno, Dr. Joseph. *The Toxic Solution: How Hidden Poisons in the Air, Water, Food and Products We Use Are Destroying Our Health*. New York: Harper One, 2017.

Poldermans, Don. "Scientific Fraud or a Rush to Judgment?" *The American Journal of Medicine* 126, no. 4 (April 2013). https://doi.org/10.1016/j.amjmed.2012.10.018.

Pompa, Daniel. "Why You May Be Breathing in Tire Particle Waste: What to Do About It." DrPompa.com. February 12, 2020. https://drpompa.com/cellular-detox/you-may-be-breathing-in-tire-particle-waste/.

Raman, Ryan. "GMOs: Pros and Cons, Backed by Evidence." Healthline. Healthline Media, July 2, 2020. https://www.healthline.com/nutrition/gmo-pros-and-cons.

Ravishankara, A. R. (Ravi). "Is Earth's Ozone Layer Still at Risk? 5 Questions Answered." The Conversation, January 5, 2021. https://www.theconversation.com/is-earths-ozone-layer-still-at-risk-5-questions-answered-91470.

Roberts, Catherine. "Stop Eating Pesticides." *Consumer Reports*, October 2020.

Robertson, Dale. "Flint Has Clean Water Now. Why Won't People Drink It?" December 23, 2020. Politico.com. https://www.politico.com/news/magazine/2020/12/23/flint-water-crisis-2020-post-coronavirus-america-445459.

Roy, Jessica. "The Bizarre Backstory of Joni Mitchell's Chronic Illness." The Cut, April 3, 2015. https://www.thecut.com/2015/04/joni-mitchells-bizarre-medical-mystery.html.

Salter, Jim. "The Army Sprayed St. Louis with Toxic Aerosol during a Just REVEALED 1950s Test." Business Insider. Business Insider, October 4, 2012. https://www.businessinsider.com/army-sprayed-st-louis-with-toxic-dust-2012-10.

Schmidt, Dena. "Acetaminophen—the Most Dangerous Over-the-Counter Pain Reliever." NaturalHealth365, September 7, 2020. https://www.naturalhealth365.com/acetaminophen-pain-reliever-2005.html.

Shukman, David. "Plastic Particles Found in Bottled Water." BBC News. BBC, March 15, 2018. https://www.bbc.com/news/science-environment-43388870.

"Smog." ScienceDaily. Accessed July 28, 2021. https://www.sciencedaily.com/terms/smog.htm.

"Soil Pollution: Definition, Causes, Effects and Solutions." Conserve Energy Future, September 16, 2020. https://www.conserve-energy-future.com/causes-and-effects-of-soil-pollution.php.

Tachover, Dafna. "5G AIRGIG: What Is It and Should You Be Worried?" Children's Health Defense, November 25, 2020. https://childrenshealthdefense.org/news/5g-airgig-what-is-it-and-should-you-be-worried/.

Taylor, Deila. "Food with Hormones vs. Food without Hormones." Healthfully, July 2, 2012.

Thomas, William. Chemtrails Confirmed. Carson City, NV: Bridger House Publishers, 2004.

"Toxic Chemicals Found in Rainwater and Drinking Water throughout the U.S." EcoWatch, December 17, 2019.

Wagner, Paul. "Scientists Keep Warning about the Dangers of 5G; Will We Listen?" Gaia. Accessed July 26, 2021. https://www.gaia.com/article/5g-health-risks-the-war-between-technology-and-human-beings.

"Wake Up Call on Mobile Phones." New Dawn, November/December 2011.

"What You Need to Know about Chlorpyrifos." Earthjustice, May 11, 2021. https://earthjustice.org/features/what-you-need-to-know-about-chlorpyrifos.

Wilson, George C. "Army Conducted 239 Secret, Open-Air Germ Warfare Tests." The Washington Post. WP Company, March 9, 1977. https://www.washingtonpost.com/archive/politics/1977/03/09/army-conducted-239-secret-open-air-germ-warfare-tests/b17e5ee7-3006-4152-acf3-0ad163e17a22/.

Wittenstein, Vicki Oransky. *For the Good of Mankind? The Shameful History of Human Medical Experimentation.* Minneapolis, MN: Twenty-First Century Books, 2013.

INDEX

Note: (ill.) indicates photos and illustrations

Gates, Bill, 163, 164, 222, 257, 260 (ill.), 260–61
Gates, Bill, Sr., 260
Gates, Mary, 260–61
gender dysphoria, 109
genetically engineered (GE) salmon, 79
genetically modified organisms (GMOs), 148 (ill.), 148–50, 157, 158–64, 159 (ill.)
geoengineering, 49–52. *See also* chemtrails
George Eastman, USS (ship), 40, 40 (ill.)
Georgia Guidestones, 250–53, 251 (ill.)
GlaxoSmithKline, 230
glyceryl thioglycolate, 209
glycol ethers, 219
glyphosate, 143–50, 144 (ill.), 147–48, 149
Gordon, Bruce, 115
Gorvett, Zaria, 235–36
Gosse, Julie, 216
greenhouse gas removal (GGR), 50
greenhouse gases, 22–23
Gunatilake, Sarath, 148
gym mats, 219–20

H

hair dyes, 206 (ill.), 206–9
Hardell, Lennart, 72
Hardeman, Edwin, 145
herbicides, 143–50, 144 (ill.)
Hermann, Paul, 138
Hertel, Hans Ulrich, 78
heterocyclic amines (HCAs), 190
hidden agendas and conspiracies
    depopulation and eugenics, 253 (ill.), 253–57, 255 (ill.)
    eugenics, 259–61

forced sterilizations, 257–59
Georgia Guidestones, 250–53, 251 (ill.)
High-Frequency Active Aurolal Research Program (HAARP), 53 (ill.), 53–57
Hippocratic Oath, 41, 44
Hitler, Adolf, 129, 130 (ill.), 131, 259
HIV, 256
Holocaust, 254
hormone disrupters, 107–9
hormones, 199–200
Horowitz, Leonard, 32–33, 56
House of the Future (Disneyland), 153, 153 (ill.)
housing pollution, 93–94
human breast milk, microwaved, 79

I

I G Farben, 154–55
ibuprofen, 233
Impossible Burger, 202
indoor air pollution, 19–22
industrial pollution, 93
ion exchange filters, 116

J

Jayasumana, Channa, 148
Jelter, Toril H., 84
Johnson, Dewayne, 144–45
Johnson & Johnson, 229, 234

K

Kapoor, John, 230
Kennedy, Robert, Jr., 95
Kenya, forced sterilization, 258–59
kidneys, 232–33
Kimbrell, George, 163
King, David, 51–52
Knowlton, Kim, 23